The Taste of Bread

A translation of *Le Goût du Pain,
comment le préserver, comment le retrouver*

by

Raymond Calvel

Professeur honoraire de boulangerie de l'École française de Meunerie
Professeur honoraire de l'Institut de Boulangerie du Japon
Président d'honneur de l'Association des Amis du Pain français au Japon
Chevalier d'honneur de la Confrérie suisse des Chevaliers du Bon Pain

Translator
Ronald L. Wirtz, PhD
Evans, Georgia

Technical Editor
James J. MacGuire
Le Boulangerie Passe-Partout
Montreal, Quebec
Canada

Photographer
Garfield Peters

SPRINGER SCIENCE+BUSINESS MEDIA, LLC
2001

Library of Congress Cataloging-in-Publication Data

Calvel, Raymond.
[Goût du pain. eng]
The taste of bread / Raymond Calvel; translator, Ronald L. Wirtz; technical editor, James J. MacGuire; photographer, Garfield Peters.
p. cm.
Includes bibliographical references and index.
ISBN 0-8342-1646-9
1. Bread. I. Title.
TX769 .C24 2001
641.8'15—dc21
00-066352

Original edition published under the title
Le Goût du Pain
Copyright © 1990 Editions Jérôme Villette, France
ISBN 978-1-4757-6811-4   ISBN 978-1-4757-6809-1 (eBook)
DOI 10.1007/978-1-4757-6809-1

Cover photo of Professor Calvel: © Jérôme Villette

Copyright © Springer Science+Business Media New York
Originally published by Aspen Publishers, Inc. in 2001
Softcover reprint of the hardcover 1st edition 2001

Orders: (800) 638-8437
Customer Service: (800) 234-1660

About Aspen Publishers • For more than 40 years, Aspen has been a leading professional publisher in a variety of disciplines. Aspen's vast information resources are available in both print and electronic formats. We are committed to providing the highest quality information available in the most appropriate format for our customers. Visit Aspen's Internet site for more information resources, directories, articles, and a searchable version of Aspen's full catalog, including the most recent publications: **www.aspenpublishers.com**
Aspen Publishers, Inc. • The hallmark of quality in publishing
Member of the worldwide Wolters Kluwer group.

Editorial Services: Joan Sesma
Library of Congress Catalog Card Number: 00-066352
ISBN 978-1-4757-6811-4

Table of Contents

Getting to Know Professor Calvel and Bread vii
James MacGuire

Working With the Masters ix
Ronald L. Wirtz

How to Use This Book xi
James MacGuire

Preface to the French Edition xiii
Henri Nuret

Color Plates xvii

PART I—CHARACTERISTICS OF RAW MATERIALS AND DOUGH PRODUCTION 1

Chapter 1   **Flour** 3
                Type and Condition of Wheat Milled into Bread Flour 3
                Nature of Breadmaking Flour 3
                Technical Characteristics of Breadmaking Flour 11

Chapter 2   **Dough** 15
                The Composition of Dough 15
                The Influence of Processing Agents and the Use of Additives 16
                Additions to French Bread in Certain Foreign Countries 17
                The Influence of Ingredients 18

PART II—THE ROLE OF MIXING AND OF YEAST FERMENTATION IN THE CREATION OF BREAD TASTE 25

Chapter 3   **Mixing** 27
                Mixing: Dough Production and the Physicochemical Development,
                    Oxidation, and Maturation of Dough 27
                Excessive Oxidation and Its Consequences 30

Chapter 4   **Fermentation** 38
                The Role of Bread Fermentation 38
                The Influence of Different Breadmaking Methods on Taste 40
                Evolutionary Changes in the Different Breadmaking Methods 45

Chapter 5	**Organic Acids**	**49**
	The Identification of Volatile Organic Acids and Their Influence on the Taste of Bread	49
	The Relationship of Organic Acids, Mixing Intensity, Dough Oxidation Level, and Bread Production Method	50
Chapter 6	**Dough Maturation and Development**	**55**
	The Influence of Dough Maturation Level	55
	The Effects of Changes in pH and Residual Sugar Levels	56
	The Effects of Loaf Molding	57
	The Effect of Type and Degree of *Pâton* Development	60
	The Effects of Freezing Unbaked and Parbaked Loaves	61

PART III—BAKING AND KEEPING QUALITIES OF BREAD AND THEIR RELATIONSHIP TO TASTE 65

Chapter 7	**Bread Crust**	**67**
	Ovens Used in Bread Baking	67
	Formation, Coloration, and Degree of Crust Baking and Their Relationship to Bread Taste	69
	The Effects of Oven Steam on Crust Taste	73
	Flour-Dusted Breads and Crust Taste	75
	Scaling of Bread Crust	76
	Frozen Storage of Baked Bread	77
Chapter 8	**Bread Crumb**	**78**
	Formation and Baking of the Crumb	78
	Crumb Color and Cell Structure	79
Chapter 9	**Bread Staling**	**80**
	Storage and Staling Effects on Bread Taste	80
	Bread Staling and Factors that Influence It	81
	Consumption of Stale Bread	83
	Shelf Life and Taste of Industrially Produced Packaged Breads	83
	Types of Bread Spoilage	84

PART IV—TRADITIONAL AND SPECIALTY BREAD PRODUCTION 87

Chapter 10	**Basic French Bread**	**89**
	Breadmaking with *Levain* and with *Levain de Pâte*	89
	Yeast-Raised French Bread (*Pain Courant*)	94
	Rustic (Country-Style) Bread with Pure-Wheat Flour	99
Chapter 11	**Specialty Breads**	**102**
	Specialty Breads	102
	Breads for Filling or Topping	119
	Savory and Aromatic Breads	120

PART V—YEAST-RAISED SWEET DOUGH PRODUCTS, COMMON AND DIETETIC RUSKS, BREADSTICKS, CROISSANTS, PARISIAN AND REGIONAL BRIOCHES ... 129

Chapter 12 Rusks and Specialty Toasted Breads ... 131
Rusks (Biscotte Courant) ... 131
Gluten-Free Breads ... 137
Breadsticks and *Grissini* ... 137

Chapter 13 Yeast-Raised Sweet Doughs ... 141
Traditional Croissants ... 141
Chocolate-Filled Buns from Croissant Dough ... 148
Snail Rolls ... 149
Brioches ... 149

Chapter 14 Regional Brioches ... 158
Regional Brioches ... 158
Vendée-Style Brioche ... 161
Specialty Brioches ... 163
Brioche-Type Hearth Cakes from Other Lands ... 175

PART VI—NUTRITIONAL VALUE OF BREAD, BREAD AND GASTRONOMY, BREAD AND THE CONSUMER ... 181

Chapter 15 Qualities of Bread ... 183
The Nutritional Value of Bread ... 183
Caloric Content and Bioavailability ... 185
The Progressive Decline of Bread Consumption in France ... 188
Bread Made from Stone-Ground Flour ... 189
Bread and Gastronomy ... 190
Comparing Bread with Other Foods ... 192

Selected Works of Professor Raymond Calvel ... 197
Compiled by Ronald L. Wirtz

Index ... 201

Getting To Know Professor Calvel and Bread

I first met Professor Calvel during my long internship at the well-known restaurant of Charles Barrier (who is incidentally a friend and mentor of Joel Robuchon). Barrier used to make his own smoked salmon, hams, *pâtés*, etc., but his great love was bread. He had a fully professional breadmaking operation to make bread for sixty lunch-time customers, and then more bread for sixty dinner customers.

At one point there were problems with the bread, and it came as no great surprise to see Raymond Calvel arrive to look into things, since he and Barrier are great friends. The Professor took the train from Paris, asked a few appropriate questions about the formula and ingredients, had lunch in the dining room, and returned to Paris on the train. He left behind a small sheet of paper on which, in his small and extremely neat handwriting, he had written a few suggestions—increase the proportion of prefermented dough, add a bit more water, mix a little less, and ferment a bit more.

I wasn't then a baker, and the changes seemed insignificant. Nor did I realize that by applying these few simple suggestions it would be possible to produce superior bread in just four hours from start to finish. Calvel had applied a principle that can be summed up in an apparently contradictory phrase, of which he is most fond. *Panifier vite et bien*—make bread quickly, but make it well.

I was fascinated, and when the changes were tried out, just a few hours after the Professor's departure, I was amazed at the results. When I tasted that wonderful bread, I was hooked, and have been ever since.

Bakers have always been known for their desire to form friendships with their fellows and for their willingness to share. This has been especially true in my dealings with the professor's friends and admirers throughout the world, and each has endeavored to make my understanding of French baking and North American flours a bit less vague. Special thanks on *this* particular project is due to Hubert Chiron, who has been extremely generous in his assistance.

Much has happened in the world of artisan baking since 1991, when we began our efforts to see "Le Goût du Pain" appear in English. Professor Calvel's principles have lost none of their pertinence—*au contraire*. With the greater popularity of artisan breads and the growing number of persons interested in this field, they are more applicable than ever.

My greatest fortune in working on this project has been to come to know and collaborate with Ron Wirtz. At the outset, Ron was to translate and I was to look into the technical aspects of putting the recipes and techniques into a North American perspective. The boundaries blurred very quickly, and many of the footnotes are his. My thanks go here to Ron, inspired translator, valiant worker, and valued friend.

We apologize for any errors that may have eluded us, and we wish our readers every success in understanding and baking great bread.

Finally, one is tempted in situations like this to think of something cute or clever to say to one's family, when in fact it is best to stay with the script: thanks to Suzanne and Lawrence, and apologies for the time and trouble.

James MacGuire

Working with the Masters

As a young university student in the mid-1960s I had the opportunity to attend a summer course in French language and civilization at the Sorbonne. That summer changed my life in more ways than one. Not only did I fall completely in love with the French language, but I came to know and appreciate the distinctive breads of France. It was a matter of great sorrow for me to find that such bread was almost completely unobtainable in most of the United States at that time. It was a loss that I felt keenly for many years.

When I came to the American Institute of Baking in 1987 as director of what is still today the finest single library on baking technology in North America, I was elated to find a sizable collection of professional baking journals in French. Since I was also disappointed in the quality of the so-called French bread produced at the AIB, I became determined to research as much as I could about the subject. As I read through article after article in several different baking journals, it became apparent to me that many of the best technical articles and books had been written by one man—Professor Raymond Calvel.

It was unfortunate that I did not meet Professor Calvel in 1987 when he presented his second seminar on hearth breads at Kansas State University. I did receive a copy of the course notes, however, and found that they followed closely the principles that Calvel had outlined in *La Boulangerie Moderne*. I continued to read Calvel's articles as they appeared, and in a 1989 article in *Le boulanger-pâtissier* was intrigued to find his statement that the best French bread in North America was made in a small bakery called *Le Passe-Partout*, owned and operated by James MacGuire in Montreal. When I visited Montreal nearly four years later, I made a pilgrimage to the *Passe-Partout*, and discovered that Calvel had spoken the truth. I also discovered that James and I had a common wish—that is, to see "Le Goût du Pain" appear in English. It is a wish that has taken us a long while, and a great deal of diligent work, to finally bring to fruition. I am happy to note that through it all we have become—and still remain—close friends. I can also say with all sincerity that he is one of the finest, most dedicated and kindest men I have ever known.

There was a special challenge to translating "Le Goût du Pain". Although Professor Calvel is a very respected scientist and researcher, his syntax is often uniquely poetic. I sometimes found myself in awe of the Professor's ability to express very technological matters in an extremely lyrical form. In retrospect, I feel that this is fitting. Calvel is a unique combination of scientist, activist, and artist, such as only France could produce.

Through the kindness of Hubert Chiron, to whom I owe a special debt of gratitude for his assistance on this project, I did finally get to meet Professor Calvel at the Europain conference in Paris in 1996 and again in 1999. It was one of the high moments of my life to converse with a man whom I had come to respect and admire so deeply, and I feel extremely grateful to have been able to collaborate on this effort to bring his breadmaking principles to the English-speaking world. I apologize for any errors that may have crept into the translation, and to those who may be offended by my distinctively American writing style and vocabulary.

I wish to dedicate my efforts in this work to the memory of my grandfather, Ronald Louis Létourneau, who taught me to love the land and the wheat that grows upon it, and gave me such pride in my Québec heritage. Thanks to my wife, Karen, and my daughter Carolyn and foster daughter Seong Eun (Esther) Oh for their great patience.

Ronald L. Wirtz, PhD

How To Use This Book

The legendary French restaurateur Fernand Point was fond of saying that great cooking is merely a series of simple operations, but that each stage must be completed successfully. This is certainly the case with French baking. Professor Raymond Calvel is a very meticulous man, and readers would be well advised to follow his example carefully. Because hearth breads have so few ingredients (basically flour, water, leavening, and salt), and because they have been made for centuries, one is tempted to oversimplify the process. Indeed, there are talented bakers who measure in handfuls and care little for methodology, but they are extremely rare.

I assure readers that frequent reading and rereading of the facts and figures in the professor's writings haven't diminished for me the fascination and mystery that surround great bread, and they have made me a much better baker. Nothing, however, can replace hands-on experience—the privilege of seeing and feeling the texture of a properly kneaded and sufficiently fermented dough. This must be transmitted on the job, from one baker to another, just as it has been throughout history. We also have to remember that beyond the facts and figures, there is a basic truth that Professor Calvel often repeats to students and to additive manufacturers—*people are going to eat the stuff!*

Great baking starts with the proper equipment, and it is certain that a steam-equipped deck oven, in which the loaves bake directly on the oven floor, remains the ideal. Bakers who tend to be obsessed with fancy (and expensive) equipment should know that the great majority of the loaves pictured in this book were mixed in an ancient 80-quart Hobart planetary mixer and baked in a used deck oven that could be termed adequate at best. All too often, the word "equipment" tends to mean ovens and mixers and overlooks the extreme importance of an accurate scale and a good thermometer for water and dough temperature calculations.

Readers who own a copy of the original French edition will note a few minor adjustments to formulations, all of which have been reviewed or approved by Professor Calvel. Most of the formulae in this English edition have been adjusted so that the recipes in each section are based on the same weight of flour. The original baker's percentages given in the French edition have been retained.[1]

For users of the metric system, one way of understanding baker's percentages is to move the decimal point one digit to the right, and to say to oneself: "For each kilogram of flour, I will need…." This is the beauty of the metric system, and it is my sincere patriotic wish that the United States will join the rest of the world in using it sometime before the next millennium arrives. For users of the U.S. system, one might say to oneself: "For each 100 pounds of flour I will need…."[2]

The use of baker's percentages and the standardization of recipe weights makes it possible to analyze and compare recipes. For example, the similarities between *brioche* and *pain brioché* (brioche bread) are great, but one quickly notices from the recipes that *pain brioché* is much less rich than *brioche*. Although calculating baker's percentages and comparing recipes will never replace bowling or Parcheesi as favored pastimes, doing so will greatly enhance the reader's grasp of Professor Calvel's breadmaking principles.

The professor's formulae are tried and true, and neophytes especially must forgo the temptation to make changes, at least until the recipe has been mastered. It should also be borne in mind that temperatures and fermentation times are as important to a recipe's success as the ingredient quantities. Any adjustments should be fairly minor. For instance:

- Kneading times might vary slightly, but it would be a mistake to underknead. Certain of Professor Calvel's North American admirers follow his principles to an extreme. It should be remembered that each bread type has its own characteristics,

---

[1] Baker's percentages are used almost universally by professional bakers everywhere, not only in Europe and North America, but in virtually every country where bread is made by trained bakers.

[2] Weight calculations (grams to ounces) were made using a commonly available Sharp Elsi Mate EL-344G calculator.

and a good baguette *must* be relatively light and have a delicate crispy, crackling crust as well as the delicate yellowish crumb that proves kneading has been kept to reasonable levels.

- North American flours might perhaps benefit from slightly longer bulk fermentation times than those indicated, but it would not be correct to assume that if 2 hours is good, then 4 hours would be twice as good. The times indicated for the shaped loaves might prove to be a bit too long for North American flours and may be adjusted downward if necessary.
- One inevitable exception is the question of absorption of water (and other liquids). Careful readers will notice that Professor Calvel reassuringly fine-tunes the amounts of water in the various bread recipes because of the presence of preferments, etc. However, lurking beneath this beautiful logic is a minefield of varying flour absorption levels. (It is for this reason that the relatively minor readjustments to the formulae for croissants and brioche were not incorporated.) Whereas the professor might vary the hydration rate by 1% or 2% from recipe to recipe, the difference from one flour to another can easily be 3% to 4%. It has been assumed in this English version of the text that most bakers will be using flour milled from hard red winter wheat. Those who use flour from hard red spring wheat—attention, Canadians!—might well find themselves increasing the hydration.

It should be noted that doughs used for French bread will probably seem too wet to bakers who are accustomed to North American textures. This is an area where experience and hands-on learning are precious, but it is certain that neither firm doughs nor the currently fashionable superhydrated doughs will produce the type of results that would meet with the professor's approval.

Certain readers might be tempted to lose patience with Raymond Calvel's fastidiousness. On every occasion in which we have worked together, I have been convinced by the results, and at times when I have drifted I have quickly seen the error of my ways. The most important piece of advice that I can give to readers—besides reading and then rereading this book—is to look seriously into a method of calculating the water temperature to ensure that doughs will be at the proper temperature at the end of mixing. (See Chapter 3.) The baker's stress levels will become much more tolerable, the recipe schedules may be followed almost exactly, and the end results will be immeasurably improved.

James MacGuire

Preface to the French Edition

The author of the book that you are about to explore asked me to present it to you. I would like to state that even if I had not known the name of its author, I would have been able to guess it from the profound knowledge of the baker's work and the great attention to the problem of good bread that is encountered in its pages.

The detailed table of contents and preface certainly give the reader such a thorough introduction to the available material that I will limit myself to saying that, after providing an introduction to the basic ingredients that make up bread dough, he proceeds to an in-depth discussion of the different bread production methods and the various factors that influence taste in both a positive and a negative sense. He then finishes this first part with remarks on the baking process and the keeping qualities of bread.

In the second part, the reader will find precise directions on the ways to make good-quality bread, along with those methods that are most appropriate for continued use, including recipes and detailed discussions. Side by side with the critical elements, there is a presentation of the positive aspects of certain irreversible technical or social developments.

Having said that, and in spite of the fact that he is well known by the whole of the French baking community, it would seem to me to be both meaningful and pleasant to follow the author just a bit along the lengthy path that he has pursued.

As early as 1948–1950, he made contacts with both English and American bakeries.

In 1954, at the invitation of Japanese baking groups (professional bakers and public agencies) and at the suggestion of the president of the French Bakery Confederation, Mr. Laserre, Mr. Raymond Calvel agreed to travel to Japan, where he spent three months and traversed the country from the north to the south.

He returned to Japan again in 1964, at the time of the Olympic Games. That incident was the real point of departure for the large-scale production of French bread in Japan. Mr. Calvel proposed setting up a stand at the Tokyo fair, where Mr. Bigot, a French baker, was able to use a battery of French equipment and the appropriate type 55 unbleached flour to make baguettes that were the highlight of the fair.

In subsequent years, Mr. Calvel returned to Japan very often, a total of 23 times so far. On the occasion of his 20th trip, Professor Calvel was honored by the combined milling and baking industries of Japan, who presented him with an honorary diploma. Today, both Japanese bakers and those French bakers who have become established in Japan continue to follow his precepts. Their observance of his principles has permitted Professor Calvel to proclaim that "the production of French bread in Japan has become a great success, and the secret of this accomplishment is that the French bread being produced is distinctively original, authentic, and of very high quality."

As a consequence, French baking equipment and French yeast have been imported to Japan, and the success of the baguette has encouraged the progressive entry of French pastry and French cuisine into Japanese society.

It is interesting as well to note that the works of Professor Calvel have been widely translated:

La Boulangerie Française was translated into Japanese and Spanish, *Le Pain Français et les Productions Annexes* into Japanese and Portuguese (in Brazil), and *Le Pain Français* into Japanese and Italian.

Professor Calvel has voyaged to quite a number of other countries, and his travels have always been marked by practical instruction that has resulted in significant improvement of the quality of the bread. He has also been able to remark that "French bread enjoys an enormous amount of positive interest (everywhere), although it is often very poorly exploited."

After dealing at length with traditional French breads, the author dedicates several chapters to specialty breads and to yeast-leavened sweet doughs, including brioches and croissants. He also discusses the nutritional value of bread in relation to the rate of extraction, and these points are especially pertinent.

In the last chapter, he discusses consumer attitudes toward bread and also the relationship between bread and various media. In this particular instance, he demonstrates a sagacity that is beyond reproach and that strongly indicates that his talents go far beyond those of a simple lover of good food.

In other parts of the text, he goes beyond technique alone to reveal the origins of the product under discussion. In speaking of his memories of a holiday where good wine was associated with the particular cake being produced, he states: "This is a also a brioche that a wine of the last vintage compliments very nicely, both of them being crowning jewels of the process of alcoholic fermentation."

Let us go on, or rather, let us return to the discussion of the quality of bread.

Professor Calvel explains in some depth the reasons for the lessening of the quality of bread: overoxidation of the dough by overmixing and the addition of lipoxygenase through the use of fava bean or soy flour. It is thus that the carotenoid pigments of the flour are destroyed, and by the same means that the normal creamy white color of the crumb and the authentic taste of bread is denatured and degraded. These causes have been known for quite some time, and the author sets forth very simple means to remedy them.

In agreement with him, I should repeat what I wrote in 1966 in an editorial in the *Bulletin des Anciens Élèves*, Number 212: "We must restore bread's original color and taste."

Why then, have things not changed, or rather, why have they changed so little?

I am told it is because the consumer wants white bread. This is both true and false: it is true because whiteness is associated with a certain idea of purity; it is false because we should associate the creamy white tint (of good bread) with the concept of whiteness. On numerous occasions I have received clear confirmation that when consumers are presented with bread having a snowy white crumb and bread with a light creamy white crumb, they have always shown a clear preference for the latter without the slightest hesitation. You can clearly observe this fact whenever there is a baker in a given area who offers truly good bread to the consumer.

Whenever I personally find myself presented with really good quality bread—especially that made at the ENSMIC[1] in particular—I always consume at least a quarter more than usual, and quite often even more than that. I am not saying that if consumers were presented with "real bread" the per capita consumption would rise again to 300 or 400 grams per day, but I am firmly convinced that quality is the only real remedy that will be effective in correcting consumption's continual decline. In my opinion, only *gourmandise*[2] is the answer.

Do we not have the means to inform millers, bakers, and consumers to orient them toward a concept (of bread quality) that will in both the short and long term be more advantageous for them and for agricultural producers? We know that the consumer would be receptive to being helped to rediscover good bread. Why, then, should we not succeed?

The work offered by this author is a positive and very opportune response to these questions, and it gives us the means to accomplish these goals.

In conclusion, I would like to say that in the mind and soul of Raymond Calvel, bread is the end product of work, of joy, and sometimes even of poetry—of everything that extends from the germinating grain of wheat until the golden loaf emerges from the oven.

It is his very life itself!

Henri Nuret
Honorary Professor of Milling
National School of Milling and
the Cereal Industries (ENSMIC)

---

[1] The École Nationale Supérieure de Meunerie et des Industries Céréalières (ENSMIC) in Paris is one of the foremost schools of milling and baking in the world. Professor Calvel has been closely associated with this school for much of his professional life.

[2] Most French–English dictionaries define *gourmandise* as greediness or gluttony. However, in this case Nuret means a "a deep and fundamental love of fine foods."

The Taste of Bread

Color Plates

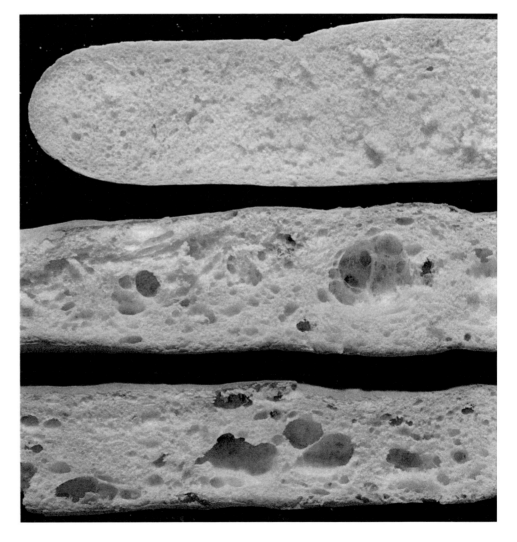

Plate 1 Cross sections, top to bottom: Intensive Mixing; Improved Mixing; Traditional Mixing

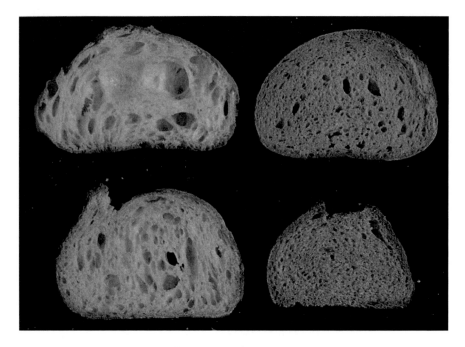

Plate 2 The baguette is not the only "French Bread". Clockwise from top right: Rye, Whole Wheat, *Levain*, Rustic.

Plate 3 Two kilogram *levain* miches. Exhibit 10–1.

Color Plates xix

Plate 4

Baguettes can be made using recipes in Exhibits 10–3 through 10–9.

Plate 5

Plate 6 *Pain de Campagne,* country-style loaves—Exhibit 11–2.

Plate 7 *La Charbonée du Boulanger* (pork shoulder braised in red wine) finds a good match with *Pain au Levain* (sourdough).

Plate 8 Buttered rye bread is a classic accompaniment to smoked fish and raw shellfish.

Plate 9 Walnut bread (Chapter 11) goes particularly well with goat cheeses.

Plate 10 Croissants and other types of viennoiseries are very much a part of the French baker's art, and are a welcome diversion for baker and customer alike.

Plate 11 Bubbles on the crust, a result of retarding raw loaves at refrigeration temperatures, are well received in North America. In France, bubbles are considered a defect.

A B C D E

Plate 12 (A) Seam placement. The seam must be on the bottom of the loaf. Often a visible seam opens up more during baking than the loaf shown here. (B) Uneven shaping. (C) Pointed ends. Professor Calvel finds the current fashion of pointed ends to be particularly irksome. Pointed ends will dry out and burn. (D) No steam. Lack of steam leads to a dull, greyish crust, and can penalize volume. (E) Too much steam. Excess steam seals the crust; preventing the slashes from opening properly and creating places where the loaf is overly dense. Rye bread, however, requires abundant steam to glaze the loaf and prevent the crust from cracking and tearing.

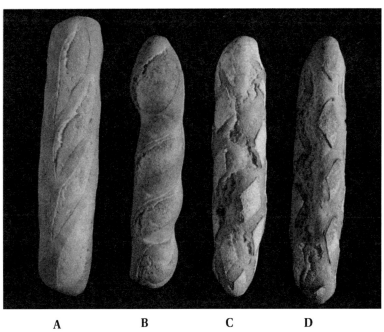

A B C D

Plate 13 (A) Overproofed loaf. An overproofed loaf flattens out. The slashes do not open properly and the texture of the crumb is penalized. (B) Improper slashing. Improper slashing has a detrimental effect on the appearance of the loaf and its oven spring. (C) Too much flour. Proper crust formation, browning, and flavor are inhibited by too much flour dusting. It is not pleasant to eat! (D) Acceptable flour. Flour on a loaf should not be the equivalent of icing on a cake! It should be remembered that slightly floured country-style loaves came about because traces of flour remained after the dough was allowed to rise in lightly floured cloth-lined "banneton" baskets. It is not, therefore, an icing-style decoration.

xxiv THE TASTE OF BREAD

      A   B   C   D   E   F   G

Plate 14 (A) Slightly underbaked. Underdone loaves do not fully benefit from the flavors created by the maillard reaction and the carmelization of the crust. (B) Proper crust color. (C) Too dark for general sale. However if a few such loaves are produced because of hot spots in the oven, they can be set aside for customers who prefer well done loaves. (D) Too dark. Burned flavors quickly mask the delicate flour and fermentation flavors. (E) Burned bottom. Uneven oven temperatures or moving loaves around as they bake can lead to bottom burning. The burned flavors migrate to the entire loaf. (F) Pale, thick crust. Hyperdiastatic flour or an overly fermented dough lacking in residual sugars will cause pale crust that is thick and dry. (G) Intensive browning. Very high oven temperature or hyperdiasticity will cause the loaves to brown too quickly, before the crust is formed.

PART I

Characteristics of Raw Materials and Dough Production

It is really no exaggeration at all to say that producing good taste in bread is a very complex problem. In spite of that, it should certainly be well understood by the baker, and by all accounts it seems to grasp the attention of the consumer to an even greater degree.

Although in principle the basic problem of bread taste is very complex, the main elements may be divided into only five key factors:

- the characteristics of raw materials and the composition of the formula or recipe
- mixing duration and intensity during dough preparation
- bread fermentation and its several intermediate stages
- dough division and formation of the *pâtons* or unbaked loaves
- baking of the dough and the baking equipment used

The discussion that follows makes no pretense of being the last word on this subject. The purpose is rather to contribute to a better understanding of the creation of bread taste through a general discussion of the topic. It is in this open and inquiring spirit that we will proceed to an examination of these elements:

1. the type and condition of the wheat milled into bread flour
2. the necessary characteristics of breadmaking flour
3. the composition or makeup of the dough
4. the level of dough oxidation achieved in the mixing process
5. the leavening method used and the fermentation that results from it
6. the potentially damaging effects of dough division and loaf forming
7. the baking of the dough
8. the influence of storage conditions on the keeping qualities of bread

Information and commentary on the principal types of bread and yeast-leavened sweet dough products will be added to the above, along with a discussion of concerns relative to the nutritional value of bread and its place in the general realm of gastronomy.

Flour

Chapter 1

- Type and condition of wheat milled into bread flour.
- Nature of breadmaking flour.
- Technical characteristics of breadmaking flour.

TYPE AND CONDITION OF WHEAT MILLED INTO BREAD FLOUR

The type and condition of the wheat milled to obtain flour for breadmaking can certainly influence the final taste of the product. First of all, wheat must be "healthy, sound and marketable," according to the time-honored French expression. To meet these basic criteria, it must not have been exposed to excessive humidity, and it must show no evidence of all the consequences that may result from such exposure, such as the odors of wet straw, mildew, or heat damage as a result of fermentation or sprouting. Since these odors are detectable in the finished flour, it is certain that they spoil the taste of bread to a greater or lesser degree. The same is true in regards to the accidental presence of insects or any foreign odors that might become apparent.

The type of wheat will also have an effect on taste, according to whether the wheat is relatively soft and starchy or hard and vitreous. A starchy wheat berry will yield a whitish flour that is generally poor in carotenoid pigments, while a hard, vitreous wheat berry, which is higher (perhaps much higher) in carotene, will result in a creamy-tinted flour, ranging from slight to marked.

This important difference will be apparent in both the color and taste of the bread, provided that the dough oxidation that occurs in mixing does not conceal or destroy it. If excessive oxidation does not occur, the crumb of the bread will be more richly cream colored, and will have a more pronounced, more distinctive, and more agreeable flavor.

The protein content of the wheat may also significantly affect the development of bread taste. Bakers will find it in their best interest to use a moderately strong flour for French bread, with the level of protein ranging from 11% to 13% (see Tables 1–1, 1–2, and 1–3); the use of flours with high levels of protein, such as those on the order of 14% to 16% found in the United States or Canada, is not advisable.[1] When very high protein flours are used, the baker is obliged to overhydrate the dough. The excessive moisture content, in combination with the high protein, will often yield bread that is mediocre in appearance, with a soft and thin crust and a tendency to become even softer through moisture migration. The crust becomes rubbery when chewed, and in combination with the high gluten content, this excessive moisture results in a lessening of desirable bread flavor and eating characteristics.

NATURE OF BREADMAKING FLOUR

Rate of extraction and ash content play a predominant role in the creation of bread taste, and we should consider the type of flour first of all. Without any other qualifying term, the word "bread" refers to the end result of flour processing. Under normal circumstances, this can only be wheat flour, and from the point of view of taste, the bread that results will have its own distinctive qualities, which are derived directly from the grain. As far as the French are concerned, the word "bread," when used alone, always means bread made from white flour milled from wheat.

Whenever the baker undertakes the use of mixed flours, as for example in the production of mixed wheat and rye bread (a mix of from 10% to 60% of rye flour and 90% to 40% of wheat flour) or of common rye bread (a mix of from 65% to 100% of rye flour and of 35% to 0% wheat flour), taste combinations that are very different from one another will be obtained, depending on the unique nature and the proportions of the two flours that are used together.

Furthermore, whenever one mixes other flours, those derived from either cereal grains or legumes, with one or the other of the two flours noted above (wheat and rye flours), the changes produced generally result in the unique taste profiles of the latter being spoiled. It is not only the origin of the flour that is important, however. The extraction rate, that is, the quantity of flour obtained from the milling of 100 kg of

Table 1–1 Protein and Ash Content in the United States and France

Protein Content								Ash Content					
U.S. 14% Moisture	France	U.S. 14% Moisture	France	U.S. 14% Moisture	France	U.S. 14% Moisture	France	U.S. 14% Moisture	France	U.S. 14% Moisture	France	U.S. 14% Moisture	France
9.1	10.83	10.6	12.62	12.1	14.40	13.6	16.19	0.40	0.48	0.55	0.65	0.70	0.83
9.2	10.95	10.7	12.74	12.2	14.52	13.7	16.31	0.41	0.49	0.56	0.67	0.71	0.85
9.3	11.07	10.8	12.86	12.3	14.64	13.8	16.43	0.42	0.50	0.57	0.68	0.72	0.86
9.4	11.19	10.9	12.98	12.4	14.76	13.9	16.55	0.43	0.51	0.58	0.69	0.73	0.87
9.5	11.31	11.0	13.10	12.5	14.88	14.0	16.67	0.44	0.52	0.59	0.70	0.74	0.88
9.6	11.43	11.1	13.21	12.6	15.00	14.1	16.79	0.45	0.54	0.60	0.71	0.75	0.89
9.7	11.55	11.2	13.33	12.7	15.12	14.2	16.90	0.46	0.55	0.61	0.73	0.76	0.90
9.8	11.67	11.3	13.45	12.8	15.24	14.3	17.02	0.47	0.56	0.62	0.74	0.77	0.92
9.9	11.79	11.4	13.57	12.9	15.36	14.4	17.14	0.48	0.57	0.63	0.75	0.78	0.93
10.0	11.90	11.5	13.69	13.0	15.48	14.5	17.26	0.49	0.58	0.64	0.76	0.79	0.94
10.1	12.02	11.6	13.81	13.1	15.60	14.6	17.38	0.50	0.60	0.65	0.77	0.80	0.95
10.2	12.4	11.7	13.93	13.2	15.71	14.7	17.50	0.51	0.61	0.66	0.79	0.81	0.96
10.3	12.26	11.8	14.05	13.3	15.83	14.8	17.62	0.52	0.62	0.67	0.80	0.82	0.98
10.4	12.38	11.9	14.17	13.4	15.95	14.9	17.74	0.53	0.63	0.68	0.81	0.84	1.00
10.5	12.50	12.0	14.29	13.5	16.07	15.0	17.86	0.54	0.64	0.69	0.82	0.85	1.01

An important technical note about protein and ash percentages. The figures given in Professor Calvel's text are expressed as a percentage of dry matter, which is customary in France. In the United States and Canada, figures are calculated on a basis of 14% flour humidity. This means that a fairly normal-seeming 11.5% protein French flour would in fact have a 9.5% protein content in North American terms, and that a high-seeming .62% ash would be .525 in North American terms.

Courtesy of National Banking Center, Minneapolis, Minnesota.

Table 1–2 Classification for Six Types of Flour in France

Classification	Ash Content as Percentage of Dry Matter	Rate of Extraction (Correlative Method)
Type 45	below 0.50	67–70
Type 55	from 0.50 to 0.60 / 0.62	75–78
Type 65	from 0.62 to 0.75	78–82
Type 80	from 0.75 to 0.90	82–85
Type 110	from 1 to 1.20	85–90
Type 150	above 1.40	90–98

Table 1–3 North American Flour Types

Flour Grade	Protein Level	Comments
Cake	7 to 8.5 protein	It has been put forth in some circles that French flours can be imitated by "cutting" the extra strength of North American bread flours with weaker cake or pastry flours. The logic of this is attractive, but it does not pan out.
Pastry	8.5 to 9.5 protein	
Hotel and Restaurant (all purpose)	10 to 11.5 protein	No North American flour is an exact equivalent of French type 55 bread flour, and bakers must look carefully for an appropriate flour and make certain adjustments (see "How to Use This Book"). Professor Calvel has had great success in North America with both "bread" flours on the lower end of the protein range and also with "all purpose" (hotel and restaurant) flours of above average strength. Significantly, many months of flour testing conducted by Didier Rosada and Tom McMahon at the National Baking Center in Minneapolis corroborates this, for 12.5% appears to be the maximum percentage of protein desirable for hearth breads. Much work remains to be done, and artisan bread movement has begun to spark an interest on the part of mills to produce appropriate flours.
Bread	11.5 to 12.2 protein	
Premium High Gluten	13.8 to 14.2 protein	The high gluten flours are too high in gluten despite Professor Calvel's mention of stronger flour for certain recipes.
Medium High Gluten	13.3 to 13.7 protein	
Strong Spring Patent	13 to 13.3 protein	
First Clear	14 plus protein	Clear flours can add strength to rye doughs when used as the wheat portion, and where their darker color is of little importance.
Whole Wheat	14 plus protein	Stone ground whole-wheat flours are of uniform granulation and contain no additives, but must be used before the wheat germ oil oxidizes and causes rancid flavors.

Flour

Figure 1-2 A Canadian Wheat Field. Courtesy of The Canadian Wheat Board, Winnipeg, Manitoba, Canada.

■

Flour is manufactured using a long series of grindings, siftings, and regrindings, gradually extracting the maximum of endosperm while eliminating the bran (Figure 1–3). Each "stream"—that is, the white flour resulting from each individual step—possesses its own characteristics. French flours are "straight process" flours, meaning that the streams are blended back together, yielding an extraction rate of 72% to 78%. This refers to the finished white flour expressed as a percentage of the whole wheat used for its manufacture.

Unlike French straight process flours, most flours encountered in the United States and Canada are admixtures of only certain flour streams, selected for strength or other desirable characteristics associated with each commerical flour category. Our investigations into this were not fruitful, revealing only that each mill seems to do things differently and that the little information made available to bakers is of no use.

Figure 1-1 A Kansas Wheatfield. Courtesy of Kansas Wheat Commission, Manhattan, Kansas

■

There is no exact North American equivalent to French flours, and bakers in some areas might be obliged to search for an appropriate substitute. The dry North American plains produce wheat with relatively high protein content. North American bakers have long believed that high protein level is directly related to flour quality. In the case of authentic French breads, this is not so. At the same time, simply choosing North American flours with identical protein content will not guarantee perfect results.

Of North American flours, Professor Calvel prefers those milled from hard red winter wheats, grown in Kansas (Figure 1–1) and other Midwestern states, because of their baking tolerance and slightly sweet flavor.

This is not to say that flours milled from hard red spring wheat from the northern United States and Canada (Figure 1–2) are entirely unsuitable, although here, the avoidance of excessive protein levels and the elusive search for the right flour, are of greater importance. The Professor has achieved very good results on numerous occasions with flours from hard red spring wheats. Indeed, Canadian flours were used to make many of the products photographed in this volume.

Figure 1-3 A Flour Mill. Courtesy of Kansas Wheat Commission, Manhattan, Kansas.

flour, also plays a determining role. The rate of extraction most commonly used in bread flour production will be discussed in some detail later in this section.

It is generally known that wheat grains are composed of a starchy kernel and of several overlapping envelopes, from the pericarp to the protein layer (Figure 1–4). The central problem of milling technology is to separate these envelopes from the kernel (or albumen) and to reduce the latter into flour.

Unless some qualifying term is used, such as the name of a particular grain or legume, this is the product that is designated under the common name of "flour." This was the flour that I myself defined in 1952, in reference to the anatomy of the wheat grain, as being made of "the maximum of the starchy albumen, with the minimum of the covering envelopes."

To be more precise, it is generally referred to as **white flour**. However, it should be added at this point

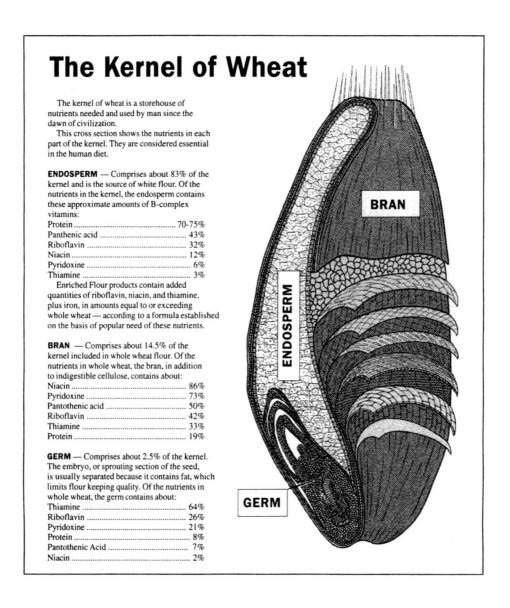

Figure 1–4 The Kernel of Wheat. *Source:* Reprinted with permission as shown in the American Institute of Baking's Science of Baking correspndence course.

The center of the wheat berry is composed of a higher percentage of starches and a lower percentage of protein. The percentage of protein increases proportionally from the center of the berry toward the exterior.

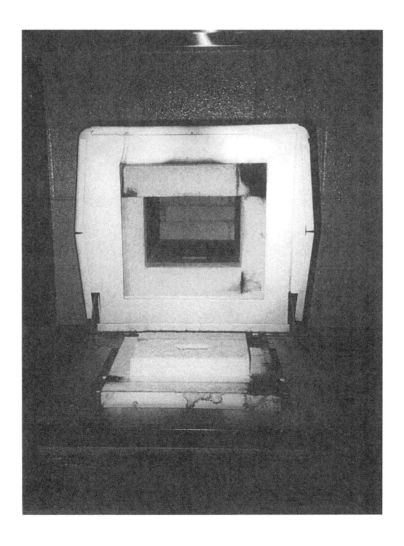

Figure 1–5 Ash Oven.

■

The mineral content of flour is determined by incinerating a laboratory sample of flour under controlled conditions and weighing the resulting ash (Figure 1–5). Because the minerals are concentrated in bran and in the outer layers of the wheat berry, the ash content is an indication of the rate of extraction. The method of expressing ash content is somewhat different in France than in English speaking countries, as noted in Table 1–1.

■

Most North American flours are composed of certain flour streams from the milling process, which are selected for strength or other desirable characteristics. This allows the miller to produce flours that are suited for the production of particular bakery products. The relatively recent development of air classification of flour particles has led to even more precise selection.

The milling process (Figure 1–6) extracts the center portions of the endosperm first and then proceeds outward toward the bran. The initial streams are therefore lower in protein and are marketed as all-purpose flours and baker's flours. Professor Calvel has achieved excellent results using regular baker's flours, and on certain occasions, all-purpose flours of above-average baking strength.

Very strong baker's flours from the middle portion of the milling process are not well suited to the artisan recipes in this volume.

Clear flours, from the end of the milling process, are darker in color because they contain bran particles. They are very strong and can be useful in the production of rye bread, where their strength is a boon and the darker color unobjectionable.

Artisan bakers should bear in mind that the majority of millers remain convinced that all bakers are interested in super-strength flour products made only from the strongest flour streams. Only in recent years have certain millers shown an interest in the production of appropriate flours for artisan hearth breads, and some very promising results have been achieved.

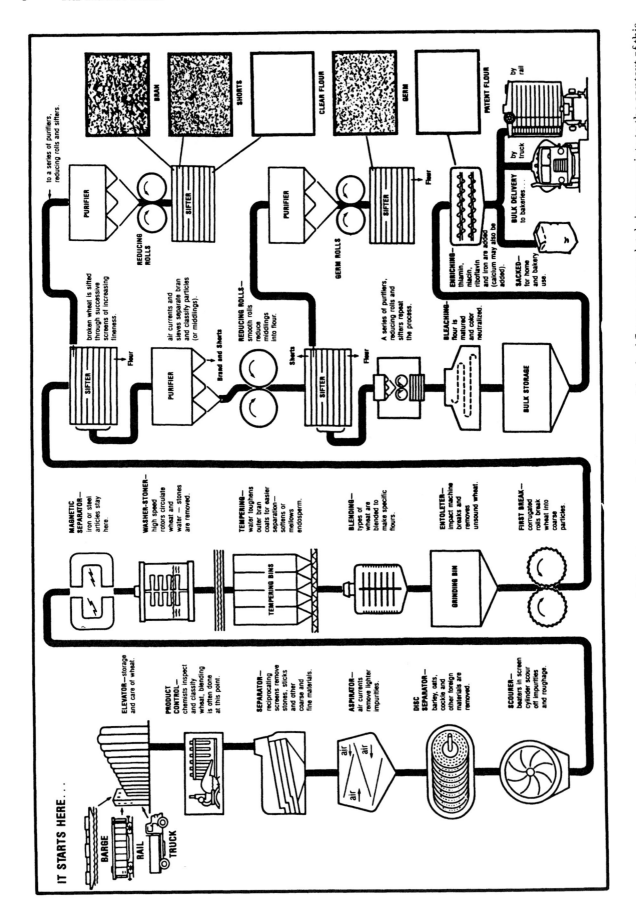

Figure 1-6 How Flour Is Milled. (Please note: Although bleaching appears in the diagram, bleached flours are completely inappropriate to the context of this book.) Courtesy of American Institute of Baking, Manhattan, Kansas.

that it is white with a more or less light creamy tint, the generic color of the starchy portion of the wheat kernel. In France, the purity level of such a flour is determined on the basis of mineral content, with an **ash level** (*see* Table 1–1) of about 0.55%. For an extraction rate from 75% to 79%, the variations in ash content are influenced by the type of wheat, its preparation, the procedures and equipment used in the mill, and the skill of the individual miller. As the ash level rises above 0.56%, it darkens the color of the flour more and more toward gray, as the flour becomes progressively richer in particles of the bran envelopes.

This flour possesses an ideal combination of qualities: it has a characteristic taste in which the flavor of wheat germ oil may be detected—a little like the flavor of hazelnuts. On the technological level, it also has an optimal combination of good plastic properties (strength, extensibility, elasticity) and fermenting properties (good dough fermentation activity, good "oven spring," that is, increase in volume of the loaves).

At a lower extraction rate, the taste of bread will still be highly distinctive, and perhaps even more tempting, but the fermentation of the dough will be slower and the oven spring less pronounced. The bread will be less voluminous and the crust will generally be thicker and less well colored. At a higher extraction rate (from 83% to 85%), the flour produced will be grayish, and the taste will change. To a greater or lesser degree (depending on the rate of extraction and other characteristics of the grain) it will take on the flavor of the bran envelopes, with an aftertaste of ash. The dough will also be different: it will be stickier, lose its elasticity, become more porous, and lose its mixing tolerance. The loaves will be less voluminous, and the color of the crumb will become grayish because of the darkening effect of the higher ash level.

As the miller begins to grind at a 98% extraction rate to make flour for so-called whole-grain breads (a flour that should really be termed "whole meal"), the dough changes even more. It is a deep grayish tan, somewhat greasy and sticky to the touch; it loses extensibility and elasticity and becomes still more porous—as the dough pieces lose their tolerance, the volume of the loaves is also reduced. They become more dense, the crumb grain becomes tighter, and the dominant taste is that of the bran layers of the grain, the taste of ash. In the same manner, bran bread, which has a slightly higher content of bran envelope particles than whole meal *pain complet*, will have an ash taste that is even more pronounced. This bread is relatively dense and low in volume but often has a lighter-colored crumb since the bran particles are larger.[2]

Whole-wheat breads and bran breads generally have a higher water content than ordinary white bread because the bran in the flour absorbs more water during dough mixing and because these more dense loaves do not lose as much water through baking. The baker must therefore be especially careful to ensure that loaves of this type are thoroughly and appropriately baked.

Thus, it may be said that the nature of the flour, its rate of extraction, and its purity are all extremely important to the overall development of bread taste.

Different Types of Flour Used in France

Regulations in France provide for the production of six types of flour (*see* Table 1–2, and *see* Table 1–3 for Information on North American flours), classified by ash content:

- Type 45 flour is used primarily in specialized products (i.e., puff pastry, brioche, and similar items) and for household uses.
- Type 55 flour is termed "breadmaking" flour. It is also used in biscuit-making (called cookies and crackers in the United States), for household uses, and in pastry-making.
- Type 65 flour is primarily used in biscuit-making. It may also be mixed with those flours used for the production of country-style breads.
- Types 80, 110, and 150 are used on a limited basis and find their primary application in the production of dark and whole meal breads.

The ash content of a flour does not have a direct relationship to the quality of the flour in terms of breadmaking value. As has been discussed, however, an under-extracted flour will produce results that are slightly inferior to a type 55 flour; furthermore, an increase of the extraction rate will affect results even more.

Wheat Grit, or Farina, Flour

Even though it is used only marginally at present, this flour should be mentioned here because of its breadmaking characteristics. It is a strong flour of superior quality, similar to those obtained from fine particles of the gluten-rich outer layer of the wheat berry in the days of stone mills.

Today, wheat grit flour is produced by specially milling strong breadmaking wheats, which have high levels of good-quality protein. This results in the production of two types of flour, type 45 and type 55 (Exhibit 1-1), which must meet the customary minimal quality standards. In actual practice, the quality characteristics of these flours are markedly higher than the minimum.

These flours may be used to produce farina bread, which was commonly made until about 1940, but which for all practical purposes has nearly disappeared today. However, farina flours are presently used, either pure or in blends, in the production of sweet yeast doughs, in general baking, and in the making of pastries such as brioches, croissants, and pan breads.

Different Types of Rye Flour

There are four types of rye flour in France. These are classified as shown in Table 1-4.

The ash content of rye flour varies over a wide range, and types 70, 85, and 130 have standards that overlap. That is largely due to a high degree of irregularity in the mineral content of the grain, and for a given type of flour it may vary according to mill grind lots, milling conditions, and the somewhat variable extraction rates used.

Table 1-4 Classification of Four Types of Rye Flours in France

Classification	Ash Content in Percentage of Dry Matter
Type 70	From 0.60 to 1%
Type 85	From 0.75 to 1.25%
Type 130	From 1.20 to 1.50%
Type 170	above 1.50%

Certain recipes in this book call for *farine de gruau*, which until the beginning of World War II was made by "skimming off" the stronger flour streams. In North American terms it corresponded to a fancy patent flour and had an extraction rate of 60% to 65%. Today's *farine de gruau* is merely a straight process flour produced from a wheat of above-average strength. In North America a regular bread flour of good strength is a suitable substitute. The use of a specialty strong flour would be going too far.

As far as flavor and aroma provided by flour to the finished bread are concerned, it seems that type 130 produces the best taste. This flour has an ash content between 1.20 and 1.30.

The three types of North American rye flour are shown in Table 1-5.

Table 1-5 Classification of Three Types of Rye Flour in North America

North American Rye Flour Types (All on 14% moisture basis)

Flour Type	Ash Content
White Rye Flour	.55 to .65%
Medium Rye Flour	.65 to 1.0%
Dark Rye Flour	1.0 to 2.0%

Professor Calvel recommends the use of medium rye flour, which has an extraction rate of approximately 80%, for the rye bread recipes in Chapter 11. Please note that there is further discussion of North American rye flours in the rye bread recipe section, Chapter 11.

Exhibit 1-1 Alveograms and Characteristics of French High-Strength Breadmaking Flour

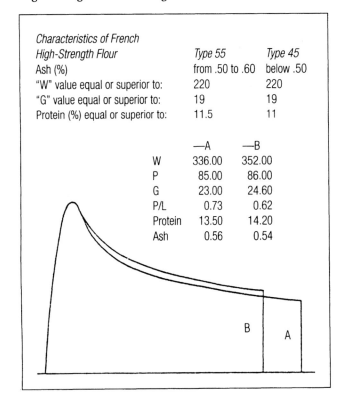

TECHNICAL CHARACTERISTICS OF BREADMAKING FLOUR

The baking quality of a flour determines its technological properties. The baking value of a flour is an indication of its ability to produce attractive and good-tasting bread under optimal production conditions and with the best yield. The quality of a flour may only be determined by test baking. However, it may be indirectly determined by examining two groups of factors:

- The baking quality of a flour is related to the quality and quantity of proteins and of the gluten that derives from them.
- The fermentation quality of the dough depends especially on its diastatic power and its fermentable sugar content.

The Baking Quality of a Flour

In France, this particular quality is determined through the use of the Chopin alveograph,[3] by means of the calculation of the "W" value (Figure 1-7). This figure results from finding the degree of work deformation of a sample of dough at a constant level of hydration. A dough sample in the form of a bubble is subjected to air pressure, and the deformation of this bubble is recorded as a curve, the **alveogram**. The derived W number represents the work of deformation expressed in thousandths of ergs per gram of dough.

The alveogram, which is the curve recorded with this instrument (see Exhibit 1-1 and Exhibit 1-2), thus furnishes the W value, which is in relation to the dough surface; the pressure or (P) factor, which is the height of the curve; and the inflation index (G), which is shown in relation to (L), the length of the curve.

These factors provide information on

- the strength of the flour;
- the cohesiveness (P) and the water absorption ability of the flour; and
- the extensibility (G) and its ability to enable greater loaf volume.

It is necessary to add that there must be a good P/L relationship or ratio expressed by the configuration of the curve.

The Diastatic Power of a Flour

This quality of flour is generally measured by the **falling number** time index as shown by the Hagberg apparatus (Figure 1-8), which determines the starch gel forming properties of the flour.

It is advisable to add a test of flour acidity to the two quality tests outlined in Exhibit 1-3. This flour pH or acidity test allows a determination of its state of conservation. It is important for the baker to know

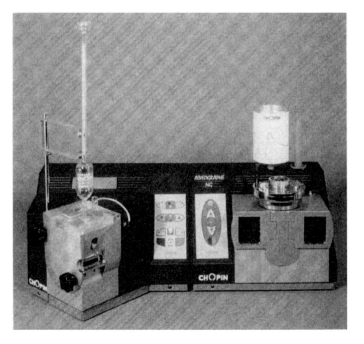

The quality of the protein in a given flour is of far greater importance than the actual percentage. Proteins give strength and structure to the dough, and in French breads must possess the qualities necessary to permit unmolded loaves of bread, which bake directly on the hearth, to hold their shape and have the right degree of oven spring. The Chopin alveograph tests a sample of unyeasted dough by inflating it until it bursts. The data recorded in this test indicate the flour's potential for breadmaking. Although the alveograph is widely used in France and many other countries around the world, it is relatively little known in North America in comparison to other types of dough testing equipment.

Figure 1-7 The Chopin Alveograph. Courtesy of Seedura Equipment Company, Chicago, IL.

Figure 1-8 Falling Number Apparatus.

Flour and water are heated until the starches gelatinize, and the viscosity of the result is measured. Low enzyme activity results in a thick gel (high falling number), and high enzyme activity leads to greater breakdown of starches and a less viscous gel (lower falling number, and in more extreme cases, hyperdiasticity).

---

the moisture content of the flour, which influences its keeping quality and its yield in bread.

Bearing in mind these different tests, the flour that would be appropriate for proper production of French bread would correspond to the quality criteria shown in Exhibit 1-2.

As regards the W value, some much stronger flours that have been milled from strong wheats can give excellent results provided that the protein level is equal to or less than 13.5%, as in the case of flours with a W value of between 300 and 350.

Evolution of the Strength of Breadmaking Flours

Finally, one cannot discuss the technological characteristics of breadmaking flour without speaking of the evolution of the quality of wheats and of flour over the past 50 years. Until around 1935–1940 the quality (of both wheat and flour) was poor.

As an example, consider flour strength as expressed by the W value, which measures the strength of the physical properties of doughs. In the past this measurement reached an average of 90 in the Parisian region and reached a level of only 70, with difficulty, in the provinces. Today, these numbers have more than doubled: in a normal year, Chopin W values are found at a minimum of 140 and are virtually always between 160 and 180.

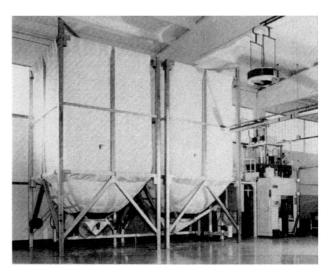

Figure 1-9 Automatic Flour Silo. Courtesy of Technosilos/F.B.M. Baking Machines, Inc., Cranbury, New Jersey.

Larger bakeries are often equipped with flour silos, which save space and facilitate flour handling. The more recent cloth designs are less prone to mildew and other problems.

Unbleached flours must be aged for 2 to 3 weeks following milling to achieve optimal performance. It should also be noted that unbleached flours carry the inherent danger of viable insect larvae, therefore requiring strict cleaning and hygiene procedures.

Exhibit 1–2 Characteristics of French Flour, U.S. Winter Wheat Flour, and Canadian Spring Wheat Flour

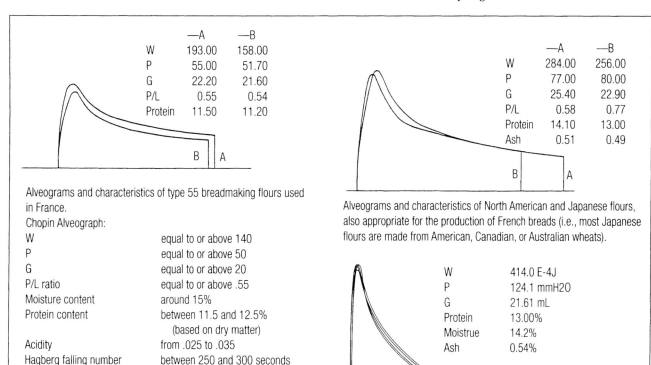

	—A	—B
W	193.00	158.00
P	55.00	51.70
G	22.20	21.60
P/L	0.55	0.54
Protein	11.50	11.20

Alveograms and characteristics of type 55 breadmaking flours used in France.
Chopin Alveograph:

W	equal to or above 140
P	equal to or above 50
G	equal to or above 20
P/L ratio	equal to or above .55
Moisture content	around 15%
Protein content	between 11.5 and 12.5% (based on dry matter)
Acidity	from .025 to .035
Hagberg falling number	between 250 and 300 seconds

	—A	—B
W	284.00	256.00
P	77.00	80.00
G	25.40	22.90
P/L	0.58	0.77
Protein	14.10	13.00
Ash	0.51	0.49

Alveograms and characteristics of North American and Japanese flours, also appropriate for the production of French breads (i.e., most Japanese flours are made from American, Canadian, or Australian wheats).

W	414.0 E-4J
P	124.1 mmH2O
G	21.61 mL
Protein	13.00%
Moistrue	14.2%
Ash	0.54%

Right: Alveogram and characteristics of Canadian Wheat Flour. Canadian flour alveograph courtesy of Robin Hood Mutifoods Inc., Etobiocoke, Ontario, Canada.

Exhibit 1–3 Tests of Flour Diastasticity

Hypodiastasticity is manifested in sluggish fermentation, lower loaf volume, and a pale, thick crust. The addition of small amounts of enzyme-active malt products is the usual remedy. Readers will notice that the bread formulae involving preferments contain more malt than straight dough recipes. This is to compensate for diasticity "eaten up" during long fermentations.

Hyperdiastasticity causes slack, sticky doughs that ferment too quickly, leading to collapse of the raw loaves and a thin, soft crust. This condition is caused by sprouting of wheat under damp conditions before harvest or by high levels of starch damage in the flour.

The harder a wheat is, the greater the starch damage it suffers during milling. It is curious to note that many of the characteristcs attributed to North American flours in general are also characteristics associated with starch damage:

- doughs become sticky toward the end of mixing
- shaped loaves ferment rapidly and tend to flatten-out
- slashes do not open properly during baking
- crusts soften shortly after baking.

This matter remains largely unexplored but informal conversations with Hubert Chiron of the INRA (French National Agricultural Research Institute) and other French bread experts conjecture that insufficient tempering of the wheat prior to milling, and rough milling practices may play a key role in this matter. Given these tendencies, and having in mind that many North American flours already contain malt products, the quantities of malt indicated in the formulae may prove unnecessary or even deleterious. Test bakes with these factors in mind will quickly clarify the situation.

Before World War II, doughs were easier to mix and to develop, and the dough became a coherent, smooth, and homogeneous mass very rapidly. In order to reach the same stage of development, doughs made from the wheats available today require a progressively greater degree of mixing and much more mechanical work input.

In the first instance (pre-1939 flours), contact of the dough with atmospheric oxygen was minimal, and the dough retained a creamy color; in the second (with present-day flours), the dough requires much more mixing. If the baker does not exercise great care, the dough will become somewhat more oxidized and a bit whiter and will lose flavor in proportion to the amount of mechanical energy required to mix it. As will be mentioned in later chapters, certain additives, combined with excessive mixing, encourage and accelerate this phenomenon of oxidation.

All of these tests, facts, and figures are extremely useful for establishing parameters for flour selection. However, nothing can replace test baking of flour samples as the absolute criterion of flour quality.

In the long run, it is best for bakers to deal with millers who are capable of providing a consistent product even though these flours might prove to be slightly more expensive. Neophyte bakers in particular would be well advised to seek out flours that are known to be appropriate for hearth breads. This will eliminate one variable from a long list of headaches.

NOTES

1. In spite of this general statement, Professor Calvel has shown numerous times that it is certainly possible to make very good bread with higher-protein North American flours, both U.S. hard red winter wheat flour and the even higher-protein Canadian hard spring wheat flour. Bread production based on these wheats certainly requires proper attention to flour blending in order to achieve proper baking characteristics, and also makes systematic bake testing of flour lots absolutely necessary. For a discussion of Professor Calvel's pioneering efforts in this area, see the articles printed in *Boulanger-Pâtissier* and listed in the bibliography, especially those that describe his work at Kansas State University in the 1980s and in Japan.

    Most flour used in Japan since the 1960s has been a blend of various grain lots from the United States, Canada, Australia, France, and lesser sources. As Professor Calvel has noted, the production of French bread from flours typically available in Japan may pose some interesting technical problems, but it is certainly achievable. This may be because Japanese milling practices are reportedly more gentle than those used in North America.

2. It has been written that ash content figures are of more interest to millers than to bakers on a daily basis, since ash content can be used as an indication that milling equipment is properly adjusted.

3. This instrument is used somewhat rarely in North America. Several other tests are widely used to test wheat flour properties. For a discussion of wheat flour testing methods in North America, see the numerous papers and test methods available from the American Association of Cereal Chemists (AACC). Extensive references are available on the AACC web site, currently located at http://www.scisoc.org/aacc/.

Dough

CHAPTER 2

- The composition of dough.
- The influence of processing agents and the use of additives.
- Additions to French bread in certain foreign countries.
- The influence of ingredients.

THE COMPOSITION OF DOUGH

It is certain that the taste of bread cannot be dissociated from the composition or makeup of the dough. For French bread, the composition is simple: a dough based on 100 kg flour will contain around 60 L water, 2 kg salt, and 2 kg yeast. Depending on the particular case, a relatively small number of additives may be added, including

- ascorbic acid, incorporated either into the flour at the mill at a rate of 20 to 60 mg per kilogram of flour or into the dough by the baker at the rate of 20 to 80 mg per kilogram of flour, with a maximum permissible use level of 300 mg per kilogram of flour;
- lecithin, at a rate of 100 to 200 g per 100 kg of flour, with a maximum permissible use level of 300 g per 100 kg;
- cereal amylases, in the form of malted flour or malt extract, up to a maximum permissible use level of 300 g per 100 kg of flour[1];
- fungal amylases, at levels ranging from 10 to 30 g per 100 kg of flour, with the level of concentration being stated by the supplier to the user; and
- faba (fava) bean flour, which is mixed with bread flour at the mill at levels ranging from 0.6% to 0.8% (with a maximum permissible level of 2.0%).

Figure 2–1 Weighing Equipment. Courtesy of Edward Behr.

Careful use of precise weighing equipment (Figure 2–1) leads to predictable results.

Since the early 1980s, faba bean flour may be replaced by 0.2% to 0.3% of soybean flour.

Other additives may also be used in the production of unsweetened specialty breads, such as

- calcium propionate or propionic acid, used to retard the growth of molds in sliced and packaged breads. The permissible use level is 0.5% maximum of finished product weight, which equates to approximately 0.7% to 0.8% of the additive based on total flour weight;
- citric acid, limited in use to the production of mixed rye–wheat bread or whole rye bread[2] (citric acid may be used up to a level of 500 g per 100 kg of flour); and
- powdered wheat gluten (obtained during the production of wheat starch, as a result of the starch–gluten separation stage). There is no limitation on the use of wheat gluten as an ingredient added to bread flour for the production of common bread, specialty breads, and high protein (dietary) breads.

The largest part of bread production in France—common white hearth breads—results from the use of standard type 55 flour. This flour type by itself would seem to have only a limited influence on the creation of bread taste. Within the broad general classification of French breads, the type of flour should be a matter for consideration only in the case of specialty product manufacture. However, the production level of these items is relatively minor in comparison to total bread production.

In any case, a lack of vitreous qualities in the wheat berry, and an ash level approaching the maximum of 0.60%, may sometimes play a limiting role in the development of desirable flavor characteristics.

THE INFLUENCE OF PROCESSING AGENTS AND THE USE OF ADDITIVES

Ascorbic Acid

Ascorbic acid (or vitamin C) reinforces the physical properties of dough and hastens its maturation (Exhibit 2–1). Furthermore, it increases the forming and handling tolerance of the unbaked dough pieces and promotes the production of larger volume loaves since it allows them to be baked at a higher proof level. It should be pointed out that ascorbic acid is destroyed by heat in the course of the baking process, and that no trace of it remains in the finished bread.

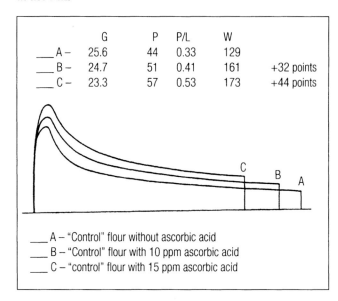

Exhibit 2–1 Effect of the Addition of Ascorbic Acid to Flour at the Mill

	G	P	P/L	W	
A –	25.6	44	0.33	129	
B –	24.7	51	0.41	161	+32 points
C –	23.3	57	0.53	173	+44 points

A – "Control" flour without ascorbic acid
B – "Control" flour with 10 ppm ascorbic acid
C – "control" flour with 15 ppm ascorbic acid

Do not be alarmed by the amounts of ascorbic acid, which are expressed in milligrams, or in the form of even more alarming parts per million. In many areas, ascorbic acid is available from bakery supply houses in easy-to-use forms. Follow manufacturers' guidelines to obtain the amounts indicated, and ensure that the product in question does not contain any undesirable ingredients. It is also possible to add 1 g of ascorbic acid crystals to 1 kg of flour or to 1 L of water. Each gram or milliliter of the resulting mixture would then correspond to 1 mg of ascorbic acid. Be sure that the ascorbic acid crystals being used are a free-running dry powder that has been protected from light. Ascorbic acid mixed with water should be discarded at the end of the production day because it oxidizes rather quickly.

Contrary to the influence with which it has often been credited, ascorbic acid has practically no influence at all on the development of bread flavor.

Since it can accelerate the maturation of the dough and thus permits a reduction in the length of fermentation, ascorbic acid can indirectly limit the formation of organic acids, which contribute to the development of the taste of bread. Thus, the improper use of ascorbic acid may somewhat handicap the full development of bread taste.

Lecithin

Lecithin is a fatty material produced industrially from soybean germ. It is also found in egg yolk. Lecithin has emulsifying properties and functions as a lubricating agent during dough formation. In addition, it has antioxidant properties. Even at a usage level of 0.1% to 0.15%, it exhibits a tendency to slow dough oxidation and the resultant bleaching effect and thus to protect the flavor complex of bread from being spoiled.[3]

Amylases

Both cereal and fungal amylases are used to correct hypodiastatic flours and to reestablish the amylolytic balance that is appropriate to proper bread fermentation. When fungal amylases are used judiciously, their secondary effects add to the tolerance of the dough and contribute to greater volume of the finished loaves.

The addition of amylases to hypodiastatic flours, whether directly into the flour or at the dough mixing stage, plays a modest but positive role in the development of the taste of bread: amylases permit the proper rising of the dough throughout fermentation and allow normal increase in volume of the loaves (oven spring) during baking. Furthermore, amylases aid in the development of optimal crust color, which is an important contributor to taste, and they also improve the shelf life of bread.

Cereal amylases result from the malting of barley or wheat and may be found in the forms of malt flour, dry malt extracts, or malt syrups. The use of these malt products helps to enrich the aroma of bread. Malt flour is generally added to bread flour at the mill; malt extracts are more rarely added by the baker during dough mixing.

Fungal amylase is a natural substance, obtained from the culture of *Aspergillus oryzae* or *Aspergillus niger*. Completely pure and highly active forms of the enzymes are mixed with a starch carrier, then blended with bread flour at the mill. Fungal amylases have practically no direct influence on the development of bread flavor.

Calcium Propionate and Propionic Acid

These substances are used in France and in numerous other countries for their antifungal properties. Their function is to inhibit and delay the development of mold, notably in the commercial production of specialty breads and sweetened-dough bakery products (for example, white pan breads), for which the salable life may extend over a fortnight. These breads are generally sliced before being cooled and packaged, with an intended shelf life of around 2 weeks.

These preservatives give the breads both a taste and an odor that are highly disagreeable and atypical of bread. This acidic and slightly putrid character is striking, especially when the package is first opened, since these odors are then released from the package headspace where they have become concentrated.

Citric Acid

The use of citric acid is limited to the production of rye breads or mixed-flour breads that are based primarily on rye flour. French regulations specify that it should be used in a premix with pure rye flour or premixed with rye blend flours that are to be used in the production of rye breads.

The use of citric acid allows the production of rye bread doughs that are slightly less sticky, easier to handle, and more cohesive. The end result is more regularly rounded breads, with a less-sticky crumb and better shelf life. Broadly speaking, there is considerable advantage to the use of citric acid in such breads, especially as regards the development of flavor.

Bean Flour

The presence of bean flour in breadmaking flour has an extremely detrimental effect on the development of bread flavor whenever intensive mixing is used, as will be discussed later in this text.

ADDITIONS TO FRENCH BREAD IN CERTAIN FOREIGN COUNTRIES

The preceding discussion of the influence of dough composition and the effect of additives on the taste of French bread does not cover all the irregularities one finds when discussing French bread worldwide. It is not enough to add that it is forbidden, whenever one makes such bread—real French bread—to use denatured flours, or to add fats, sugars, milk powder, or any products of another type that would detract from its original, distinctive character.

French food legislation goes to greater lengths to guarantee the authenticity of foods than most other countries. A 1993 law intended to protect artisan bakers against the onslaught of industrial bakers defines *pain maison* as bread made entirely in the bakery from start to finish, using the acceptable ingredients as discussed in this chapter. *Pain tradition* goes still further, excluding the use of ascorbic acid and lecithin, but allowing faba or soy flours because they have been used traditionally. From a historical viewpoint this may make some sense, but considering the dangers of oxidation, the end result could be the opposite of that which is desired.

Whenever one speaks of French bread beyond the borders of the Hexagon (i.e., France), there are other irregularities:

- The flour is sometimes bleached at the mill by treatment with benzoyl peroxide,[4] or even treated with an addition of potassium bromate, as is done in the United States, England, and certain other English-speaking countries.[5] In both of these instances, and especially in the case of benzoyl peroxide, the gustatory properties of the flour are profoundly changed for the worse, and the taste greatly diminished.[6]
- In many countries the dough used for so-called French bread is enriched with fats. This is often true in Belgium, the United States, Latin America, Italy, and the Scandinavian countries.
- In many of these same countries the dough often contains additives of the mono- and diglyceride type or various fatty acids, as is the case in Belgium, Spain, Mexico, Argentina, and Brazil. Potassium bromate is often added as well, sometimes at excessive levels, as in the United States, England, and Argentina.[7] Such treatments and additions affect the taste of "French" bread, degrading and denaturing it, making it atypical and often even unrecognizable. Such bread has only the name and the vague form of French bread, the shape of the "stick," as Finnish bakers say. This does not exclude the sad fact that bread is becoming insidiously more soft and pastry-like as a result of intensive mixing and the use of lipoxygenase. As we will see further along in the text, these practices have been combined with other assaults on bread to become nearly as damaging in France as they are elsewhere.

The flavor of French bread results from the production process, and after fermentation, from the baking of a dough that is elegantly simple in composition. This simplicity is the basis of its quality and its delicacy; but because of this simplicity, the flavor of bread is an extremely fragile attribute. It cannot resist the abuses inherent in the use of an improper flour and the addition of incorrect ingredients, adjuvants, or additives, all of which tend to unbalance the delicate aromatic equilibrium that is the true taste of bread.

THE INFLUENCE OF INGREDIENTS

The Influence of Water

Water plays a basic role in the breadmaking process: by hydrating the flour, it humidifies the starch granules and proteins. After these proteins are transformed into gluten, they serve as the linking agent to enclose the starch granules in the gluten matrix, re-

---

Flour additives encountered in English-speaking countries

Acceptable for use in production of authentic French breads:

- vitamins (added to flour as an obligatory enrichment in many areas)—thiamin mononitrite, riboflavin, niacin, folic acid, iron
- malt products (aids fermentation and crust color)—malted barley flour, amylase
- ascorbic acid (used as a dough conditioner)—preferably added to flour by the baker, in Professor Calvel's prescribed amounts, rather than at the flour mill.

Unacceptable for use in production of authentic French breads:

- benzoyl peroxide (used as a flour bleaching agent)—destroys the carotenoid pigments that Professor Calvel considers to be flavor carriers.
- potassium bromate (an oxidizing agent)—banned in many countries and localities because of health concerns
- azodicarbonomide (an oxidizing agent)
- fava (faba) bean or soybean flour
- dough conditioners added directly by the baker or at the flour mill as part of the milling process—sometimes added by the baker during dough mixing. They do not generally contain unacceptable ingredients, but in any case dough conditioners are unnecessary in the context of this book and should be avoided in the production of traditional breads.

---

sulting in the creation of a dough mass. Furthermore, water creates the humid environment appropriate to the development of enzymatic activity and bread fermentation. Whatever its source, this water must be potable. With very rare exceptions, it almost never has any appreciable effect on the taste of bread.

However, when the water is obtained from a municipal water supply, the purification treatment used should result in suitable water conditions. If water purification includes the addition of chlorine, it must be added in the proper amount. This is important, since an excess might inhibit yeast activity and thus spoil the taste of the bread. Furthermore, it is necessary to determine the mineral composition of the water and ensure that the degree of hardness is around 25. For very practical processing reasons, this is often more critical when using spring water.[8]

In the case of water that is too hard, it may be necessary to soften it somewhat, which will simplify its use at all stages of processing. One must consider the desired physical properties of the dough: water that is too soft reduces cohesiveness, while water that is too hard reduces extensibility. In either of these cases, the quality of the water may indirectly and negatively affect the taste of the bread.[9]

The Influence of Salt

Salt has a very important role in the creation of bread taste, both directly and indirectly. It has been only since the end of the 18th century, however, that it has been in common use in the production of bread in France. In a general sense, its positive influence is apparent throughout the entire production process. During mixing, it reinforces the plastic properties of the dough and improves both its cohesiveness and elasticity.

Having understood salt's role as an antioxidant, too many bakers since the 1960s have adopted the practice of delaying the addition of fine table salt—which dissolves quickly—until 5 minutes before the end of mixing. This is done with the intensive mixing method in order to encourage the maximum levels of oxidation and bleaching. This delayed salt method tends to facilitate the forming of gluten bonds during dough formation and results in a slight improvement in dough strength. However, there is also such a great decline in the quality of the taste of bread produced by this method that it might be considered a general disaster. This practice has the effect of "washing out" the dough.

It would be wise, whatever the length and intensity of the mixing method used, to add the salt at the start of the mixing stage, or at the latest, 3 minutes after beginning the operation, just as was done up until the 1960s.[10]

In addition, it should also be noted that

- salt also facilitates the development of crust color, which tends to improve the taste. In the absence of salt, the crust will remain pale.
- salt improves the flavor and odor of bread.
- due to its hygroscopic properties, salt influences the shelf life of bread.
- in dry weather, salt maintains keeping qualities by slowing down desiccation or drying.
- in damp weather, salt causes a decline in shelf life by increasing crust softening. Thus, it is advisable in the first instance to protect the bread from drafts, and in the second to allow some exposure to air.

The amount of salt added to the dough has been affected by intensive mixing and the washing effect that results from it. The unnatural, white, and insipid appearance of the bread crumb has caused bakers to increase the salt content in an effort to find a means of compensating for lost color and taste, but at the risk of committing a nutritional blunder. Before 1956–1960, the salt content of most bread was from 1.8% to a maximum of 2.0%.[11] At the present time, the usual amount of salt added is 2.2% and sometimes even higher. The growth in percentage is about 15%, which is considerable.

The medical profession certainly suggests that a lower salt content is healthier. It would seem reasonable, in order to no longer wash out the dough (and to comply with modern medical guidelines) to return to the former 1.8% added salt level.

The Influence of Yeast

Baker's Yeast

Baker's yeast is a fermentation agent that belongs to the *Saccharomyces cerevisiae* species and is actually a member of the mushroom family. It is a biological leavening agent that possesses the basic attributes of all living things, which are respiration and reproduction. Its role in breadmaking is to create the panary fermentation process, which produces carbon dioxide gas. At the same time, yeast is the "artisan" who creates the internal cellular structure of dough and causes it to rise. Yeast fermentation is also involved in the maturation of the dough and contributes greatly to the production of bread taste.

Baker's yeast is a natural biological product that is produced industrially. In France, it is grown on a molasses substrate. This is a byproduct of sugar production that still contains between 40% and 50% sugar and constitutes the basic food necessary to yeast production. Traditional yeast, known as "fresh yeast," is a somewhat firm and homogeneous paste-like product, with a creamy or ivory tint. It is composed of round or oval cells about 1/100 of a millimeter in diameter, weighing about 8 to 10 billion to the gram. It is packaged in rectangular blocks of 500 g each. Fresh yeast has a very high water content—around 70%—and only a limited shelf life: 10 to 15 days at 15°C (59°F), or 30 days at 0°C (32°F). When it is stored at 20°C (68°F), it is advisable to use it as soon as possible.

This commercially produced yeast (Figure 2–2) is generally of uniformly good quality. The taste of bread made through the use of this "biological" baker's yeast is relatively low in acid and is dominated by the taste and aroma derived from wheat flour. These elements are combined with flavor components produced by alcoholic fermentation of the dough and as a result of the baking process.[12] In French bread, the taste of yeast is not discernible until the usage level reaches 2.5%. Beyond that level its presence becomes more

Figure 2–2 Yeast Production. Courtesy of Lallemend Inc., Montreal, Canada.

and more noticeable as the usage level increases. Without being really disagreeable in itself, this is an atypical taste that seems undesirable in bread.

Dry Yeasts

Two types of dry yeast are obtained from fresh yeast by subjecting it to low-temperature drying processes. In comparison to fresh yeast, dried yeasts have the advantages of being less bulky for shipping and handling and also have a very long shelf life when hermetically packaged.

The older of these two types is available in the form of small granules containing only 7% humidity. The yeast cells, which are in a latent state, must be reactivated in advance in order to be used as an active leavening agent. This is done by rehydrating the yeast in five times its weight of water at a temperature of 38°C (100.4°F), along with a little sugar. After 15 to 20 minutes of rest, the reconstituted yeast may be put to use.

This yeast gives excellent results when used at a level equivalent to half that of fresh yeast. However, at higher usage levels, it has the same unfortunate tendency as fresh yeast to transmit to the dough—and thus to the bread—a more pronounced yeast taste, which is atypical of good bread as well as unpleasant.

The second dry yeast type, more recently developed, is a so-called "instant" yeast with a 4% moisture content that is available in the form of small rod-shaped pieces, similar to vermicelli in shape. When premixed with flour during the mixing stage, this type has the advantage of functioning much as fresh yeast. When used at a level of one-third that of fresh yeast, it also gives excellent results. It is more subtle than fresh yeast as far as taste is concerned, and especially more so than granulated dry yeast. Even at a usage level of 2% or above, its aroma is not evident. This instant yeast seems to accord more importance to the influence of the taste of wheat flour (natural and pure), to alcoholic fermentation, and to the effects of baking, which are particularly beneficial to the taste of bread.

In summary, bread fermentation resulting from the use of yeast at a reasonable level yields doughs that mented cultures based on yeasts and bacteria that are found in the atmosphere and that might be termed "ambient yeasts."[13] In the opinion of some specialists, these yeasts belong to the *Saccharomyces minor* family. According to the majority of other scientists, however, these cultures or sponges are made up of rather poorly defined fermenting agents and are sometimes referred to under the general term of "wild yeasts." A culture of these yeasts may form naturally on a piece of dough exposed to the open air, provided that surrounding environment remains temperate or even slightly warm and humid. This will result in the beginnings of an alcoholic fermentation, which after a period of several days will be followed by an acid fermentation. This method of leavening doughs made from cereals suitable for breadmaking dates from far back in the night of time: its place of appearance is generally considered to be on the banks of the Nile, about the time of Moses.

According to the legend, a piece of dough made from flour and water is supposed to have been forgotten through carelessness, and the dough was naturally inoculated by wild yeasts from the surrounding environment. Because of the effects of alcoholic fermentation, the dough would have increased in volume, and the idea of mixing it into an inert dough must have sprung into the mind of some anonymous housewife. It is thus that breadmaking fermentation was born.

The warm and humid climate of the Nile Valley[14] appears to constitute an ideal location for natural inoculation, and this inclines me to accept as likely that which the legend presents.

One can verify this hypothesis even today, without any problem. A present-day example of this phenomenon may be observed in summer, during stormy or warm and humid weather, in the laboratory of mills equipped with a Chopin alveograph. During the trials with this apparatus, the scraps of dough used in the course of tests are thrown into a wastecan. (This dough contains no added yeast.) Whenever one forgets to empty the scrap dough from the wastecan at the end of the week, it will be full on Monday morning, even though it was only one-third full on Friday. A natural inoculation by ambient yeasts has taken place, followed by production of carbon dioxide gas and rising of the dough.[15] It would suffice to take 500 g of dough from this beginning state and cultivate it by adding flour and water several times to produce, 48 hours later, a lightly acetic fermentation. This is the "building" of a natural sponge, suitable for leavening a dough with which to produce a naturally leavened bread.

Figure 2–3 Commercially Available Culture. Courtesy of Lesaffre Ingredients, Marcq-en-Baroeul, France.

A recent development in both Europe and North America has been the introduction of commercially available powdered sourdough cultures (Figure 2–3). There are two basic types: *active cultures* are prepared from the powdered base one day in advance of use, are used just as "home grown" cultures, and yield acceptable results. Dried *deactivated cultures* do not ferment and constitute an effort to give sourdough flavor and acidity to yeast doughs. Connoisseurs of real sourdough are not deceived.

Still another development is the introduction of liquid sourdough equipment, which is generally used in conjunction with commercial active cultures. These are practical and show promise but could lead to the use of liquid levain in almost every recipe (including those where sourdough is not a usual ingredient) with monotonous results.

---

are characterized by active fermentation. This process produces bread with a characteristically delicate taste, which owes as much to the influence of flour and to the baking process as to the process of alcoholic fermentation itself.

The Influence of Starters

Fermentation and breadmaking with a natural sponge starter or *levain* have as a beginning fer-

This example is especially significant in that there is very little probability or risk of a dough being inoculated by a culture of natural baker's yeast. Furthermore, the atmosphere of a flour mill is highly charged with ambient yeasts and bacteria suitable for inoculation of a dough based on wheat or rye flour because of the presence of wheat and milling dusts.

Breadmaking from a natural sponge (often called a sourdough in the United States) persisted in France and in much of the rest of the world until the dawn of the 20th century. This was (and is) a production method by which the baker makes up sponges daily. These cultures are used in turn to leaven the doughs with which the baker carries out the day's production.

A piece of dough drawn from a batch made the previous day—known in France under the term *chef* (head or chief)—was cultivated from one to three times, according to custom and the needs of the day, during the interval that separated the end of the previous day's production from that of the next. This involved the makeup of successive sponges, the last of which were used to leaven the dough used to actually produce bread. This was a slow fermentation process, and the rather dense bread that resulted had a distinctive taste. Without being disagreeable, the flavor was somewhat sharp, with pronounced acetic origins.

Toward the end of the 19th century, and just at the beginning of the 20th, city bakers especially began to add a small amount of baker's yeast to the dough in order to slightly speed up the fermentation process. The resulting bread gained in volume, while losing a bit of its acid taste. This was the era of production of large loaves, of which certain types were elongated in form. However, the majority were rounded and known as *miches* or *tourtes*.

Alongside this production based on natural sponges developed a system in which baker's yeast took on a larger role. This method often replaced the older one with a production method that took the name of *levain de pâte*, or preleavened dough. This method is sometimes encountered even today in certain rural areas. It still requires the daily preparation of a sponge of a slightly acidic nature from a chef.

As we have noted, natural sponges produce taste and aroma profiles that are predominately acetic. This same type of flavor, but at a notably lower level, is also found in those breads that result from the use of the *levain de pâte* method.

In times past this acetic flavor was accompanied by other aromas that made a more complex and agreeable composite. However, since the appearance of intensive mixing, the excessive oxidation that results is transformed into a bleaching effect that denatures both color and taste. The taste of these breads is thus dominated by an aggressive acetic or vinegar character, which makes them frankly disagreeable. They are perhaps even more distasteful than those breads leavened with baker's yeast that may have undergone the same intensive mixing treatment. These problems will be further discussed in later chapters.

NOTES

1. Many North American flours already contain malt products, which are customarily added at the mill. Bakers are advised to verify whether their flour does or does not contain malt and make adjustments as necessary.
2. Citric acid is of greater benefit to yeast raised rye breads than to sourdough ryes.
3. Lecithin is used primarily in large industrialized bakeries, rather than in artisan production.
4. Professor Calvel says *azote*, which is the inert gas nitrogen. However, he actually means benzoyl peroxide, which is used in the bleaching of white bread flours. Chlorine, another bleaching agent, is used primarily for cake flours in common practice.
5. Since this was originally written, many of the English-speaking countries have banned the use of potassium bromate outright or placed restrictions on its use, as have a number of U.S. states.
6. Azodicarbonomide is added to certain North American flours and should also be avoided in the production of French bread. It permits the elimination of the bulk fermentation, with predictably disastrous effect on the taste and other desirable characteristics of French bread.
7. See above note. The use of potassium bromate is also banned throughout Canada.
8. It is also important in many parts of the United States when using water pumped from private or municipal deep wells. Much of the United States has ground water that exceeds 240 ppm of dissolved calcium carbonate. See: *Water Atlas of the United States* by Thomas Geraghty, David Miller, Fritz Van Der Leeden and Gary Troise: Water Information Center, © 1973. *The Reference Source* annual (formerly *Bakers' Reference Source*) published by Sosland Publishing Company, Kansas City, MO, regularly includes information on water types, buffering, and how to adjust pH and TTA of water.
9. Water that is too soft also makes the dough very sticky and difficult to process. This was one of the reasons for the development of the so-called mineral yeast foods, which were originally standardized mineral packages in a starch carrier.
10. This section was edited to clearly express Professor Calvel's antipathy to the delayed salt method. A more literal translation

might be subject to misunderstanding, since Professor Calvel does not make his point until the end of the section. A direct translation might be: "However, it should be pointed out that in France, since the 1960s, the salt is generally incorporated 5 minutes before the end of the mixing stage whenever processing involves intensive mixing. Having understood that salt plays an antioxidant role, the baker today uses fine table salt and delays its addition to encourage the maximum level of oxidation and bleaching. In addition, it should also be pointed out that the absence of salt tends to facilitate the forming of gluten bonds at the beginning of dough formation. This delayed salt method results in a slight improvement in dough strength, but there is also such a great decline in the quality in the taste of bread produced by this method that it might be considered a general disaster. This effect has in the past led me to refer to it as a practice which most certainly has the effect of 'washing out' the dough." A cursory reading of the paragraph might lead the reader to think that Professor Calvel favors the delayed salt method, whereas in actual fact he is adamantly opposed to it.

11. In the recipes throughout this book, Professor Calvel has generally used the higher 2% level.

12. Flavor-producing results of the baking process include the caramelization of crust sugars and the Maillard browning reaction (a browning of the protein in flour due to heat and enzymatic action).

13. These ambient yeasts are found not only in the atmosphere but on the surface of grain, and thus are found in flour itself.

14. Professor Calvel almost certainly means that the relative humidity is elevated because of the presence of the river.

15. Ambient yeasts include both airborne and surface-borne microflora as noted above. Some of this spontaneous fermentation is certainly due to yeasts already present in the flour, rather than inoculation from the air. It should be noted that the dough used in the Chopin alveograph tests do not contain any added yeast but may contain ambient microflora of many types.

PART II

The Role of Mixing and of Yeast Fermentation in the Creation of Bread Taste

Mixing

Chapter 3

- Mixing: dough production and the physicochemical development, oxidation, and maturation of dough.
- Excessive oxidation and its consequences.

MIXING: DOUGH PRODUCTION AND THE PHYSICOCHEMICAL DEVELOPMENT, OXIDATION, AND MATURATION OF DOUGH

Mixing is the basic operation in dough production. Its primary role is to combine the individual ingredients that make up the dough and then to ensure the input of a sufficient amount of mechanical work (energy) to produce a smooth, cohesive, and homogeneous dough. (This is a dough that pulls away from the walls of the mixing bowl and is easily detached from the hands.)

Bearing in mind the physical characteristics of the flour, the dough should exhibit plastic or rheological characteristics that are suited to the final product into which it is to be made. It should have appropriate elasticity, cohesiveness, extensibility, and receptivity to forming and sufficient gluten film impermeability to avoid the loss of fermentation gases.

It should be noted in particular that the constituents of the dough whose characteristics most actively contribute to its makeup are water and the insoluble proteins in wheat flour known as gluten—and gluten is found almost exclusively in wheat flour.

Water is added in accordance with the characteristics of the flour and as a function of the desired plasticity of the dough, that is, to achieve needed handling or machining characteristics, degree of dough development, and level of proof.

It should be noted in passing that the correct water temperature is easy to determine and is used in turn to regulate dough temperature.[1] Dough temperature in turn plays a preeminent role in determining the level of fermentation activity. Furthermore, dough temperature has an important effect on the oxidation of dough during mixing. Given that the average mixing temperature varies between 24 and 25°C (75.2°F and 77°F), it should be apparent that oxidation, dough bleaching, and deterioration of bread taste will become more pronounced at dough mixing temperatures of between 26 and 27°C (78.8°F and 80.6°F), while these same effects will be moderated, and bread flavor improved, by lower dough mixing temperatures, in the 22 to 23°C (71.6°F to 73.4°F) range (Exhibit 3-1).

Figure 3-1 Automatic Water Cooler. Courtesy of F.B.M. Baking Machines, Inc., Cranbury, New Jersey.

■

The calculation of the temperature of the water to be used in a dough is perhaps the most important decision a baker makes from one batch of dough to another. Adjusting the water temperature is the only practical means of achieving proper dough temperature at the end of dough mixing.

The primary importance of water in the dough is to ensure the formation of two types of links: those among starch grains and those among gluten particles. Insoluble proteins become hydrated as they come into contact with water, combine to form gluten,

Exhibit 3–1 Controlling Dough Temperature with Ice

The use of crushed ice to regulate dough temperature is a time-honored practice and is very useful for bakers who do not have water-jacketed or refrigerated mixer bowls. This method is especially appreciated in the summer, when available water supplies may not be as cool as desired.

Two different formulas are commonly used, depending on whether or not U.S. or metric weights and measures are used in the bakery.

The formula for U.S. standards is as follows:

$$\text{Weight of ice} = \frac{\text{weight of water (tap water temperature} - \text{required water temperature)}}{144}$$

Procedure:

(1) subtract the required water temperature from the temperature of the available tap water. This determines the number of degrees each pound of water must be cooled.

(2) Multiply the number of degrees each pound of water must be cooled by the total recipe water weight.

(3) Divide the total degrees the water must be cooled by 144 (it takes 144 British Thermal Units to melt 1 pound of ice). This gives the pounds of ice required.

Example: If 12 pounds of water are required for a given mix and the temperature of the tap water is 60°F, while the calculated desired water temperature is 45°F, the weight of ice and water required would be 1.25 pounds and 10.75 pounds respectively. Using the formula:

$$\text{Weight of ice} = \frac{12\,(60-45)}{144} \text{ or 1.25 pounds. Weight of water} = 12 - 1.25 = 10.75 \text{ pounds.}$$

The resulting weight of ice must be deducted from the weight of the total water in the recipe to adjust for the addition of ice.

To calculate ice requirements in metric units, use this formula:

$$\text{Weight of ice} = \frac{\text{weight of H}_2\text{O (Kg)} \times (\text{temp. tap H}_2\text{O in °C} - \text{required H}_2\text{O temp in °C})}{\text{Tap H}_2\text{O in °C} + 80°C}$$

Example: If 40 kg of water are required for a given mix and the temperature of the tap water is 20°C, while the calculated water temperature is 10°C, the weight of ice and water required would be 4 kg and 36 kg respectively. Using the formula:

$$\text{Weight of ice} = \frac{40 \text{ kg} \times (20°C - 10°C)}{20°C + 80°C} \text{ or 4 kg. Weight of water} = 40 \text{ kg} - 4 \text{ kg} = 36 \text{ kg}$$

Courtesy of American Institute of Baking, Manhattan, Kansas.

---

and create the dough structure and plasticity that is found only with wheat flour (Exhibit 3–2).

Because of the effects of mixing, the proteins enrobe the starch granules and form an enveloping network of gluten fibrils. This network undergoes progressive development under the effects of mixing, ultimately resulting in the formation of gluten protein films, which become progressively thinner and more continuous. The formation of this fine, thin film is facilitated by the links that form between the proteins and the wheat germ oils found in the flour, as shown by the increasingly smooth and cohesive character of the dough. It should be understood that the closeness of these complex links is directly related to the level of mixing achieved, which greatly influences the properties of resulting dough.

There is thus a physicochemical process that accompanies the purely mechanical action of dough mixing. Under the combined effects of increased mixing speed and cumulative mixing time (Table 3–1), this process brings on the phenomenon of **dough oxidation**. Within limits, oxidation has the short-term ef-

Exhibit 3-2 Dough Types and Degree of Hydration

> Until fairly recent times, it had been generally accepted that three types of bread doughs were produced in France, based on the degree of hydration. However, in a recent article in the bulletin of the *Amicale*, Professor Calvel added a new category to this traditional list, the *pâte mouille*, or superhydrated dough.* Although the best-known example of this type of product is the Italian *ciabatta*, superhydrated dough products have also been made for generations in France in the Ségalas area around Rodez and in the Lodève region, south of Larzac.
>
> The combination of high water content and gentle treatment of the doughs produces loaves with large irregular holes or cells in the interior and good keeping qualities. One problem that may occur with this type of water-rich dough is rapid crust softening, and a gummy or rubbery crust and crumb is often the result. The high water content also makes it difficult to form an adequate gluten network, and this may lead to a tendency to overmixing and excessive oxidation of the dough.
>
> | Dough type | French hydration rate | Approximate North American hydration rate |
> |---|---|---|
> | stiff doughs = *pâtes fermes* | 55% | 60% |
> | medium doughs = *pâtes bâtardes* | 60%–62% | 66% |
> | soft or loose doughs = *pâtes douces* | 65%–66% | 69% |
> | superhydrated doughs = *pâtes mouilles* | 70% | 73%–76% |
>
> *For more information on this type of dough, see: "Un quatrième type de pâte – la pâte surhydratée ou la pattemouille" in *Fidèles au bon pain: Bulletin de liaison de l'Amicale des Anciens Élèves et des Amis du Professeur CALVEL*. No. 17 (1998).

fects of increasing the strength of the dough, hastening its physical development, and reducing the length of time necessary for maturation of the dough.

The process of **dough maturation** is thus linked, through the degree of development of the gluten network, to the level of oxidation reached in the mixing process. Furthermore, as the mechanical development and oxidation of the dough increase together, the process of dough maturation will accelerate over time.

Conversely, whenever the mechanical development of the gluten network and oxidation diminish as a result of a decrease in the level of mixing, it becomes necessary to lengthen the maturation period to obtain adequate development of the gluten network. Just as in the old days of slow mixing, this maturation is accomplished by alcoholic fermentation and the dough rise effect that results from it. However, the mechanical development of a dough is not necessarily followed by its oxidation.

When the speed of the mixer is greatly accelerated, the length of time that the dough remains in contact with atmospheric oxygen is reduced to only a few minutes. If the degree of air exposure of the dough is further reduced by the arrangement of the bowl or of the mixer blades, the gluten network will rapidly reach a state of optimal development, while dough oxidation will be very limited.

All other conditions being equal, this is what happens with some high-speed mixers, in which the optimal development of the dough is achieved just at the end of mixing. With this method, oxidation is practically nonexistent, and at the end of mixing the dough

Table 3-1 Approximate Mixing Times for Each of 3 Mixing Speeds (Various Types of Doughs)

	Oblique Mixer			Spiral Mixer		
Speed	1st	2nd	3rd	1st	2nd	3rd
Sourdough	5	12	17	3	4	7
Prefermented sponge	5	10	15	3	3.5	6.5
Conventional yeast, slow speed	3 + 12	0	15	3 + 3	0	6
Improved straight dough	5	10	15	3	4	7
Poolish	4	10	14	3	3.5	6.5
Prefermented dough	4	8	12	3	2.5	5.5
Rustic	4	8	12	3	2.5	5.5
Gruau	4	9	13	3	3	6
Country-style	4	8	12	3	2.5	5.5
Rye	5	5	10	3	1.5	4.5
Whole wheat	3	12	15	3	4	7
Bran	3	12	15	3	4	7
Soya	4	10	14	3	3.5	6.5

maintains nearly the same color and pigmentation it had at the beginning. If there is a positive result for the aroma of bread that results from the distinctive nature of this high speed production method, there is also a disadvantage—a notable lack of dough strength. This is relatively easy to remedy by a longer first fermentation or a higher dough temperature, or as a last resort a larger dose of additives. However, the strength of the dough absolutely must be corrected if the appropriate degree of maturation is to be reached, culminating in the production of quality bread.

Mixing is thus a good deal more complicated than it appears (Figure 3–2). It is necessary to know at the same time both its requirements and its effects, and to carefully regulate the process in reference to the equipment in use.

Oxidation has a determining influence on the degree of dough maturation during mixing, as will be discussed further in the chapter on bread fermentation.

When mixing intensity is limited (i.e., slow mixing with moderate mechanical working of the dough), the maturation is practically nonexistent. Alcoholic fermentation will be needed to complete the maturation, so bakers should schedule a relatively long primary fermentation stage, as was done before the 1955–1960 period, to achieve a natural maturation of the dough.

Since the development of the gluten film is also quite limited with the slower mixing method, the volume of the loaves will be slightly below average. However, they will be properly raised, and their flavor and taste will be quite distinctive. They will be appetizing and good-tasting, with an extended shelf life.

On the other hand, a very high level of oxidation is achieved, and dough maturation is greatly accelerated, whenever:

- intensive mixing methods are used,
- bean flour or other oxidizing additives are present,
- salt addition is delayed until late in the mixing process, and
- a large portion of the dough is continually in contact with atmospheric oxygen.

The use of these practices (along with the improper use of ascorbic acid) produces an artificial maturation of the dough.

This artificial maturation results in an almost complete elimination of the primary fermentation or **pointage** and deprives the dough of the organic acids it would otherwise produce. It is the absence of these acids that diminishes the taste, flavor, and shelf life of bread.

However, an even more serious matter is that the uncontrolled oxidation that occurs during intensive mixing causes bleaching of the dough. Further along in the breadmaking process, this will mean that the crumb portion of the loaf will be bleached, resulting in the lessening and general deterioration of the taste of the bread.

In general, whenever maximum development of the gluten film and the accompanying hyperoxidation of the dough result in very high-volume loaves, the desirable gustatory qualities are proportionally reduced.

EXCESSIVE OXIDATION AND ITS CONSEQUENCES

Since 1958–1960, breadmaking methods that cause excessive oxidation of both dough and finished bread have been practiced in France, and as a result there has been a closely linked and very significant degradation of both bread taste and quality. From that time, the majority of French bakers acquired two-speed mixers and adopted the practice of intensive mixing. This resulted in increases in both the rotational speed of the mixing vanes and the duration of the mixing stage itself.

It was also discovered during the 1960s that because of soy flour's high lipoxygenase content, the addition of 1% bean flour could greatly increase the amount of dough oxidation that occurred during intensified mixing.

Faba (fava) bean flour,[2] with a higher sugar level than wheat flour, was an adjuvant that had been known for a long time in France as a means of correcting certain quality faults. It was used until the late 1960s to correct the problem of bread that was difficult to bring to the proper crust color, or *pains dur à la couleur*. This was because the wheat was hypodiastatic in two of every three harvest years, and the flour produced from it yielded light-colored bread with a hard, thick crust and often a reduced volume.[3] During that period bean flour was added at a minimum rate of 2%, with the maximum authorized use level of 5% being used only from time to time. With the advent of intensive mixing, however, its use became systematic. The maximum authorized use level was lowered to 2%, with the practical use level being 1%, which progressively declined to between 0.6% and 0.8%.

In dough production, the amount of mechanical work and oxidation used should be in close accord with actual mixing requirements. Although it is wise to markedly reduce the amount of mechanical work

Dough Autolysis: A Calvel Discovery

Autolysis is the slow-speed premixing of the flour and water in a recipe (excluding all the other ingredients), followed by a rest period. The other ingredients are added when mixing is recommenced. Premixing and autolysis times appear in the recipes in other chapters.

During experiments in 1974, Professor Calvel discovered that the rest period improves the links between starch, gluten, and water, and notably improves the extensibility of the dough. As a result, when mixing is restarted, the dough forms a mass and reaches a smooth state more quickly. Autolysis reduces the total mixing time (and therefore the dough's oxidation) by approximately 15%, facilitates the molding of unbaked loaves, and produces bread with more volume, better cell structure, and a more supple crumb. Although the use of autolysis is advantageous in the production of most types of bread, including regular French bread, white pan sandwich bread and sweet yeast doughs, it is especially valuable in the production of natural *levain* leavened breads.

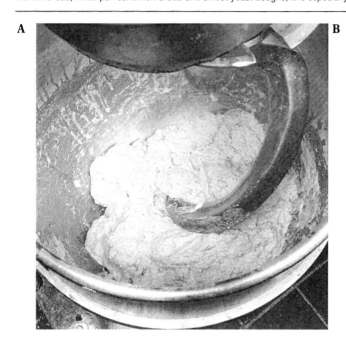

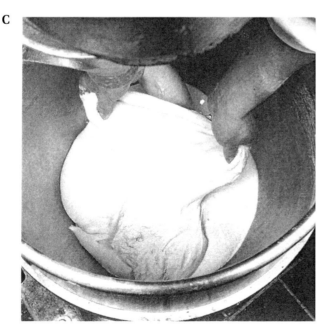

Figure 3–2 Mixing Dough. (A) At the start of mixing the dough is coarse and badly knit; (B) semi-kneaded dough; (C) fully kneaded dough.

Proper mixing involves sufficient formation of the gluten network to form a smooth, homogeneous, stretchy dough, while limiting oxidation to a minimum to avoid loss of flavor. Mixing and the maturing of the dough are mutually dependent. For this reason, the discussion of mixing continues in Chapter 4.

input, correct mixing practice must certainly include careful attention to the dough's technological needs. This is certainly true in the case of doughs made from pure wheat flours, which have undergone a very considerable increase in strength over the past 40 years.

Proper attention to technical mixing requirements permits the baker to produce doughs that are coherent, smooth, and extensible, and that have good gas retention properties in spite of the much greater gluten strength. This in turn allows the production of loaves with an appropriate degree of maturation and good oven spring. The finished bread is light and well developed, with a fine, crisp crust and a crumb that is well structured, creamy white in color, and appetizing and agreeable to eat (Figure 3–3).

This end result is possible only if dough oxidation is not excessive and does not destroy the laboriously created equilibrium discussed here. Unfortunately, such destruction is precisely what has happened since the appearance of intensive mixing.

From 1957 to 1962, during which time the addition of bean flour was neither systematic nor generally practiced, oxidation was more limited and the use of longer fermentation periods meant that dough maturation proceeded more slowly. During that time, bakers were able to produce a quality of bread that captured the hearts and the pocketbooks of the majority of consumers.

However, the goal of the great majority of professional bakers at that time was to produce white bread that was very light and high in volume. On a human level, this was a way for the average baker to prove that he was capable of adapting to progress and of mastering a new technique—only the positive side of which was seen at that time. Equipment dealers—mixer dealers especially—hawked their two-speed mixers, and within the space of a year or two, the mixing equipment of the entire industry was completely replaced with mixers for making the whitest possible bread (Figures 3–4, 3–5, and 3–6).

Summary of Mixing Methods

Traditional. Mixing is done in slow speed only, and the low level of physical dough development requires a relatively long first fermentation to compensate. It yields excellent flavors, but the resulting loaves are denser than what is seen in current practice. Professor Calvel discusses this method at length in Chapter 4. The only formula calling for this method is shown in Exhibit 10–3.

Intensive. This is truly the antithesis of the traditional method. Prolonged high-speed mixing permits the quasi-elimination of the first fermentation, irreparably penalizing the flavor, texture, and keeping qualities. The delicate flavors of the flour's carotenoid pigments are lost through oxidation. Professor Calvel was one of the first to decry this method in the 1960's. A skeleton formula, included as an indication of what not to do to make good bread, appears in Exhibit 10–4.

Improved. The majority of recipes in this book use this method which was devised as an improvement on the disastrous effects of intensive kneading. Mixing in second speed is just sufficient to yield relatively light loaves with crispy, crackly crusts while seeking to keep oxidation (and its attendant loss of flavors) to a minimum. When used with the recipes involving prefermented dough (or other yeast-based preferments) truly superior results can be produced relatively quickly. The mixing times in the recipes are for oblique mixers. Equivalent mixing times for spiral mixers appear in Table 3–1.

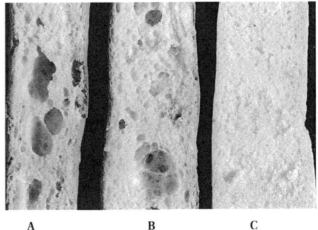

Figure 3–3 (A) Traditional; (B) Improved; (C) Intensive.

Mixing   33

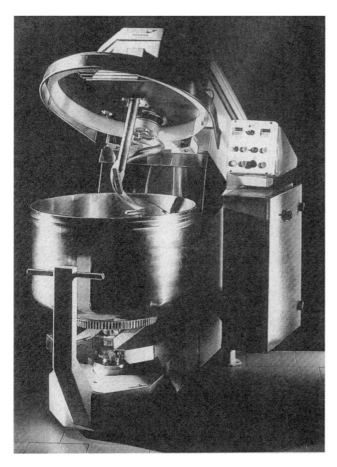

Figure 3–4 Spiral Mixers Are Effective and Limit Oxidation. Courtesy of F.B.M. Baking Machines, Inc., Cranbury, New Jersey.

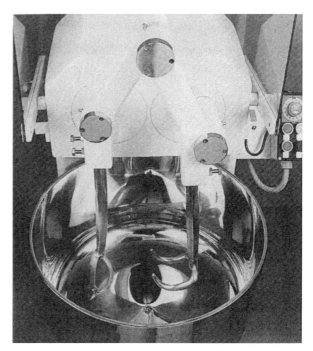

Figure 3–5 Dual-arm Artofex-type Mixers Treat Doughs Very Gently. Courtesy of F.B.M. Baking Machines, Inc., Cranbury, New Jersey.

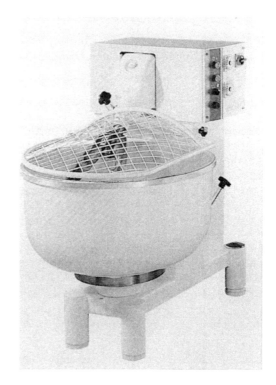

Figure 3–6 Oblique Mixers Should Be Used with Care Because They Tend To Oxidize the Dough. Courtesy of F.B.M. Baking Machines, Inc., Cranbury, New Jersey.

Millers also offered flours for white bread—the whitest bread—which had simply been improved with an added 1% of bean flour. Remember that this was the same bean flour whose virtues had been known for 80 years but whose particularly harmful qualities had been discovered only at the beginning of the 1960s.

The suppliers of additives did not stop there: ascorbic acid, carried by buffers that were as innocuous as they were different from one another, was actively promoted by zealous technical representatives. These semiskilled "sorcerer's apprentices" encouraged the use of ascorbic acid while taking advantage of it to introduce bakers to the production of an even whiter type of bread. With very few exceptions, the great majority of bakers were both converted and materially equipped to produce a new type of bread from 1962 on.

Figure 3–7 Vertical Spindle Mixer. Courtesy of Groupe Hobart Canada Équipment alimentaire, Lachine, Canada.

Vertical spindle mixers: Professor Calvel has written that most mixers, including the type shown (Figure 3–7), can be used to make great bread as long as the baker knows when it is time to push the "off" switch. The great majority of loaves pictured in this book were kneaded using an "ancient" model of this type of mixer.

Vertical spindle mixers—also called **planetary mixers**—are seldom used for bread doughs in France, but Professor Calvel finds no problem with their use. "Improved" method mixing times would be 3 minutes in the first gear of a three-speed machine, or the second gear of a four-speed machine followed by 4 to 5 minutes in the second gear of a three-speed machine, or the third gear of a four-speed machine. "Traditional" mixing would be approximately 3 minutes in first gear before the *autolyse* (or autolysis, a rest period), followed by a further 7 to 8 minutes in first gear after the *autolyse*.[4]

Even as they were taken in by the novelty of this new type of bread, many consumers very rapidly began to adopt a critical attitude toward it. The reproach was often widely made that "today's bread no longer has any taste, and it doesn't keep well either." A certain degree of suspicion began to be directed toward the baker, who was accused of bringing things to that state by using the "pastille"—a harmless tablet of compressed starch that served as a carrier for a measured amount of ascorbic acid or vitamin C. Certain consumers or journalists (infected with publishing fever) had seen it in the hands of bakers and had felt obligated to point a finger at the tablet, confusing it with the chemical substance from whence all of the trouble had come (i.e., bean flour). This was an unfortunately contrived folk tale without any basis in fact.

As already noted, ascorbic acid had nothing to do with this discoloration. The harmful whitening of the dough and of the crumb of the bread is due solely to the action of lipoxygenase (a substance that occurs naturally in bean flour). This combined with intensive mixing, the oxidizing action of atmospheric oxygen, and the practice of delaying the addition of salt until the latter part of mixing to produce a perversion and lessening of bread taste.

During this same period, the consumption of bread fell at the rate of 2.1% per year, or more than 20% in 10 years. Although other factors than the lessening of quality contributed to this situation, there is no denying that a decline in consumption did in fact occur.

What were the mechanisms by which this change in the quality characteristics of bread took place? How could such a decline be precipitated by a more intensive mechanical working of the dough and by the strong dough oxidation that resulted?

Mr. Guilbot, an administrator of the National Institute for Agronomic Research (INRA), speaking of the work of Mr. Draperon, another researcher in the same organization, discussed the action of bean flour under intensive mixing and described the phenomenon and the resulting effects as follows:

> …these effects are due to the presence of a diastase traditionally added to wheat flour; that is, lipoxygenase, an enzyme which catalyzes the oxidation of unsaturated fatty acids….
>
> This lipoxygenase is principally carried by the 0.5% to 2.0% of bean flour, traditionally added to wheat flour in France. The oxidation which it catalyzes is strongly increased during the course of intensive mixing, because there is a greater incorporation of air and formation of more extensive dough surfaces. The peroxides which are thus formed decompose, releasing volatile prod-

ucts. These products are the cause of the alteration of bread taste, because they modify the equilibrium of the substances which compose the aroma of bread, in a manner that finds very little favor with consumers.

It is this same enzyme that is the cause of the increase in bread whiteness, as a result of the oxidation of carotenoid pigments present in flour. It is also partly through the use of this enzyme as a processing aid that an increase in bread volume may be obtained. Certain oxidation products which act on the gluten facilitate the formation of the gluten network. The result is a strong gluten network that is more resistant to the passage of carbon dioxide and better able to increase loaf volume development.

An in-depth study of the basic causes of these changes, and of the mechanisms by which they operate, has resulted in the discovery of the central and multifunctional role of this enzyme. Lipoxygenase is situated at the very "crossroads" of chemical reactions in dough, and can exercise an influence not only on the taste appeal of bread, but also on its nutritional value through the loss of carotenoids and of vitamin A....[5]

It is also interesting to refer to the results of experiments conducted by the School of Baking of the Grands Moulins de Paris flour company, in the course of a joint project with the DGRST. These experiments evaluated the influence of intensive mixing in combination with bean flour and ascorbic acid on crumb color and the presence of carotenoid pigments in baked bread (Table 3-2).

It may be seen that ascorbic acid does not destroy carotenoid pigments. If the crumb appears slightly whiter to the consumer, that may be attributed to the greater degree of development of the internal crumb structure as a result of the positive strengthening effect of ascorbic acid, as is pointed out by the author of this article, Mr. Charelègue.

The preceding discussion clearly demonstrates the results of the action of bean flour on dough and finished bread. The presence of just 1% of bean flour and the lipoxygenase that accompanies it results in virtually complete destruction of carotenoid pigments, and the bleaching action is total. One last point remains to be clarified, however: the author modestly passes over in silence the considerable damage done to the taste of the bread.

Faced with such a preponderance of negative proof, why does the practice of adding bean flour to bread flour at the mill continue, even though the price of bean flour is higher today than that of wheat flour? For the miller, its use involves a certain amount of self-interest because it provides a degree of insurance in dealing with many bakers. Because of ignorance and misinformation among many bakers, a dough that stays cream colored is considered suspect and, in their estimation, is likely to produce a grayish crumb. In reality, the crumb is just lightly cream colored and is only being confused with the true color of the flour.

The weight of tradition also has an effect. There is not a single baker in France today who has not heard discussions regarding bean flour. In the dry years from the 1870s until the 1950s, when wheats were hypodiastatic or low in fermentable carbohydrates,[6] many millers improved bread flour by adding 2% to 3% of bean flour to correct crust color defects (insufficient browning)—a correction that was not detectable in any case. It was also done more willingly then, since in those days bean flour was less expensive than wheat flour. Furthermore, since there was no overmixing in those days, the risks of excessive oxidation and of bleaching were practically nonexistent.

What is the situation regarding bean flour today, and what reasons might reasonably be given for its use? For some mills it offers the advantage of simplifying the improvement of breadmaking flour through the addition of bean flour, which also serves as a carrier for cereal or fungal amylases, in addition to the lipoxygenase as a bleaching agent. These amylases are intended to reestablish the enzymatic equilibrium of the flour from these mills, just as since 1974 ascorbic acid has conveyed the commercial advantage, in terms of flour strength, of raising the W value on the Chopin alveograph test by 15% to 20%.

The use of bean flour also allows the miller to deliver young, freshly milled, "green" flours to the baker. Flour reaches its optimal level of maturation after a

Table 3-2 The Influence of Intensive Mixing in Combination with Bean Flour and Ascorbic Acid on Crumb Color and the Presence of Carotenoid Pigments in Baked Bread

Constituents	Carotenoid Pigments of the Crumb as % of Flour	Evaluation of Crumb Color at Consumption
1. Pure wheat flour control	65	—
2. + 1% bean flour	3	whiter than 1 and 3
3. + 30 ppm. ascorbic acid	69	whiter than 1

cold-weather rest period of 20 to 25 days, or after a warm-weather rest period of 15 to 18 days. This entails a bothersome requirement for flour storage, even if it is not entirely the responsibility of the mill. The presence of bean flour during mixing compensates for this lack of flour maturation by oxygenation, slightly increasing the strength of the dough, bleaching it, and partially compensating for the lack of maturation of the flour, resulting in the following advantages:

- Doughs are slightly less sticky and very slightly more elastic and tolerant.
- Bread loaves are more regular in form.
- There is a very slight increase in the volume of the loaves—from 2% to 3%, sometimes a little more, which is hardly noticeable.

It should be repeated here that the baker can obtain these results by other methods and that the use of bean flour produces a spectacular bleaching of the crumb and other harmful results with which most bakers are well acquainted. Except for bleaching, the improvements from the use of bean flour become unimportant when properly matured flour is used. They are insignificant because, with very rare exceptions, the baking quality of flour improves notably during the maturation period.

For example, the water absorption capacity of the flour increases by 1% to 2%, sometimes more. For this reason, a higher bread output will be noted for the same weight of flour, which permits alert bakers to rapidly amortize the investments necessary to increase their flour storage capacities. The production improves also: the doughs are less sticky, have a little more strength, have better tolerance, have improved quality, and demonstrate gains in regularity.

It can thus be affirmed here that the advantages provided by bean flour are for all practical purposes almost inconsequential. The phenomenon of crumb bleaching remains, and, as we know, it is a synonym for spoilage, perversion of the true nature of bread and perversion of the taste, often leaving a hint of hexanol and rancidity.

As Mr. Guilbaud points out, when the losses in vitamins and the destruction of carotenoid pigments, which act as flavor carriers and are also "A" provitamins, are added to the preceding problems, the resulting degradation is astounding, as are the indolence and passivity of the majority of professional bakers faced with these problems. What is even more astonishing is that those who nostalgically favor the use of bean flour recognize the seriousness of this degradation.

A number of tests have been conducted in an effort to avoid these negative effects. These trials involved heat deactivation of the lipoxygenase in an effort to retain only the advantages of the high levels of proteins and sugars in bean flour. At the 2% to 3% levels that were used formerly, such deactivation might sometimes have constituted a positive addition, but at the 0.7% average used today it is absolutely useless, even though the lipoxygenase is not present.

Finally, it should be noted that soybean flour, another additive that is higher in lipoxygenase than bean flour, was added during the 1980s to the choices available to the baker. At usage levels that range slightly above and below 0.2%, it can take the place of bean flour with the same unfortunate results.

A reduction in mixing has also been advised, but in comparison with the flours of the 1950s, those of today are clearly stronger. When they have been improved with the addition of ascorbic acid as well, the baker is obliged to mix more in order to obtain smooth, coherent doughs—even if the intent is not to overdo it. When using bean flour, it is thus very difficult to avoid reaching the critical zone of oxidation, spoilage, and perversion of bread taste. This is especially true whenever bakers adopt the unfortunate habit of adding salt 5 minutes before the end of the mixing period, as they often do when using conventional two-speed mixers.

To justify this practice, which in the case of intensive mixing brings on the "washing effect" and the accompanying degradation of taste, some enormous falsehoods are being shopped about: "Incorporated at the beginning of the mixing state, salt breaks the gluten links, the dough has poor strength, and the resulting bread is consequently less attractive, less regularly formed, less voluminous." (Such myths are often repeated!)

Certainly, the bread can be very slightly different. When salt is incorporated at the beginning of the mixing cycle, the oxidation of the dough will be less—and because of that, it will be less strong—and the dough will be less white. Consequently, some means will have to be found to correct this lack of strength, and several solutions exist (e.g., adequate bulk fermentation, preferments, ascorbic acid). When the corrections have been made, the resulting bread will be quite attractive, the crumb will be creamier, and it will have recovered the taste that would have been lacking—and good taste is why people eat bread.

To conclude the discussion of this important problem, it may be said that the choice of bakers in France in its widest sense may be expressed as follows:

Either they continue, through the use of bean flour and the lipoxygenase that it carries, associated with the abuse of overmixing, to wash out doughs and degrade the quality of bread, or they stop using bean flour, reduce mixing of the dough, and by appropriate means of adaptation, try to respond fully to the vocation: to produce a bread that responds to the expectations of the consumer, with constant attention to the preservation of bread's distinctive originality and to the betterment of its quality.

NOTES

1. Wayne Gisslin, in *Professional Baking*. John Wiley & Sons, 1985, p. 38, describes an extremely simple method for dough temperature calculation that is precise enough for retail and serious amateur use. It may be reduced to a formula that includes (a) needed dough temperature, (b) flour temperature, (c) room temperature, and (d) machine friction to arrive at a figure for (e) water temperature. This simplified formula uses a figure of 20°F (11°C) as an average friction factor. The formula can be stated as:

   $(a \times 3) - (b + c + d) = e$

   More precise calculation is needed for larger retail and industrial use, but this requires determination of a machine friction factor for each individual mixer used. Mr. Gisslin summarizes the procedures and information needed to calculate the machine friction factor on page 329 of the above text. This is also discussed in greater depth in the *Applied Baking Technology* correspondence course from the American Institute of Baking, 1213 Bakers Way, Manhattan, KS 66502.

2. Fava (faba) bean flour is not encountered as a flour ingredient in North America, but the effects of flour bleaching and other oxidants such as soy flour may be equally disastrous.

3. Problems of hypodiastaticity can be dealt with effectively by the addition of a very small of malt powder or malt syrup. See the references to cereal amylases and malt products in Chapter 2.

4. Planetary mixers do not develop the gluten network to the same extent as other mixer types. All bread doughs made with planetary mixers might benefit from a slightly prolonged bulk fermentation.

5. Mr. Guilbot, who was Director of Research for the INRA (France's equivalent of the U.S. Department of Agriculture), gave his opinion on Mr. Draperon's work in the "Technical Yearbooks of the INRA: 1972."

6. As happened 2 years out of 3.

Fermentation

CHAPTER 4

- The role of bread fermentation.
- The influence of different breadmaking methods on taste.
- Evolutionary changes in the different breadmaking methods.

THE ROLE OF BREAD FERMENTATION

Bread fermentation is an anaerobic alcoholic fermentation brought about through the action of fermentation agents on the sugars that are present in dough. These fermentation agents may be introduced in the following ways:

- as a natural sponge, which in bakery practice is the result of a primary culture of those agents (yeasts and bacteria) found in nature and in flour. In this instance, bakers build and cultivate their own fermenting agent. The fermentation obtained in this method is slow and may be lightly or strongly acid, with an especially acetic (vinegarlike) quality.
- by biologically pure strains of baker's yeast, which result from industrial cultivation of appropriate cultures from selected stocks that were originally chosen from brewer's yeasts. In this case, the fermenting agent is a basic material that is not the product of the baker's work. The fermentation is more active and much less acid.

The sugars involved in fermentation originate from:

- preexisting sugars such as glucose, levulose, and sucrose, which are normally present in flour at levels from 1% to 2%
- sugar generated by the breakdown of amylose and by degradation of a starch fraction by cereal amylases or other amylases present in the dough. These produce maltose in the course of fermentation.

Of these sugars, however, only glucose and levulose may be directly used by yeasts. Yeasts also provide some enzymes to the dough—invertase acts on sucrose, while maltase breaks down maltose. Other enzymes in dough hydrolyze sugars, making them fermentable and transforming them as follows:

- Sucrose is transformed by the process of *inversion*—half into glucose, half into levulose.
- Maltose is transformed into glucose.

The final step of the process is carried out by another diastase. Zymase transforms glucose and levulose into carbon dioxide gas and alcohol.

Bread fermentation, also known as panary fermentation, plays a double role during breadmaking. It ensures the production of carbon dioxide gas, which promotes the formation of alveoli and the rising of dough, as well as the oven spring, which accelerates the development of the dough pieces at the beginning of baking.

Even though all too often enough time is not allowed to accomplish its functions, fermentation also gives rise to the production of complex organic acids. These initiate a change in the physical properties of the dough, ensure its maturation, and simultaneously produce the aromas that together contribute to the development of bread taste.

Whenever this multiple activity occurs through the maturation of a dough that has been produced by slow mixing and provided with a reasonable (i.e., not excessive) amount of yeast, a relatively slow evolution of the physical properties of the dough will occur. This requires the baker to make allowances for a lengthy first fermentation period. It is this same slow fermentation that results in natural dough oxidation and that produces slow maturation of the dough, naturally encouraging the development of the unique complex of aromas that are particular to French bread.

This desired level of physical dough maturation is the result of a state of equilibrium between the opposing qualities of extensibility and cohesiveness. With rare exceptions, the degree of dough extensibility regresses steadily as fermentation proceeds, with a corresponding growth in the cohesiveness of the dough. This state of equilibrium should be reached after dough division, just as the molding of the un-

Figure 4–1 Poolish and Pâte Fermentée.

A Few Words about Dough and Leavening Systems

English-speaking bakers may be unfamiliar with the terms used for various types of French dough and leavening systems. The following four systems are *yeast doughs*.

- **Straight dough (Méthode directe).** This is the simplest fermentation method. No preferments are involved, and everything is added to the mixer at the time of mixing.
- **Sponge and dough (Levain de levure or levain-levure).** Part of the flour is used to make a fairly firm yeast culture, to which the balance of the ingredients is added at the time of mixing. Despite the confusing word *levain* in the French name of the method, this process has nothing to do with sourdough.
- **Liquid sponge and dough (Poolish).** Part of the flour and water is used to make a rather liquid yeast culture, to which the balance of the ingredients is added at the time of mixing (Figure 4–1).
- **Prefermented dough (Pâte fermentée).** A straight process dough that benefits from the addition of a portion of prefermented dough, either set aside from a previously produced batch or made especially for that purpose (Figure 4–1).[1]

The two systems below are generally called "sourdoughs" in English, although in actuality they are *much less sour* than a San Francisco–type sour.

- **Sourdough (Levain).** The use of a culture of wild yeasts and heterofermentative lactic acid bacteria, which must undergo two or three "builds" before being incorporated into a dough as the only leavening agent. Professor Calvel's method for building a levain culture appears in Chapter 10.
- **Hybrid method (Levain de pâte).** A mixed method in which a sourdough culture is used along with a small amount of baker's yeast as leavening agents for a dough. The sourdough mother culture must remain free of baker's yeast, however.

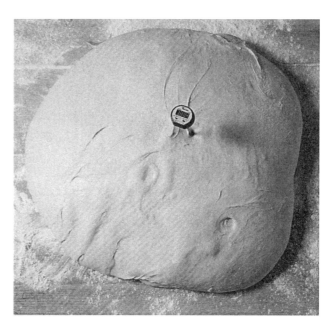

Figure 4–2 Fermenting Dough with Thermometer. Note that glass spirit or mercury thermometers should ***never*** be used for this purpose.

Proper dough temperatures and the use of a dough thermometer lead to superior and more consistent results (Figure 4–2). Always use the proper type of thermometer.[2]

---

baked loaves is being carried out, and essentially fixes the physical properties of the dough as required from this point forward in the breadmaking operation.

These physical properties include appropriate levels of cohesiveness, extensibility, and elasticity, all of which facilitate the forming or molding process and impart a good appearance and good tolerance to the formed loaves until they are placed in the oven. They also ensure good gas retention and an appropriate degree of plasticity, both of which encourage good development of the loaf during the first stages of baking.

In contrast to this natural maturation, which is accomplished without any oxidation of the dough other than that which is due to the normal progression of fermentation, the baker may bring about artificial dough maturation. This is achieved during mixing and is the consequence of a very strong oxidation resulting from overmixing, in conjunction with the action of the lipoxygenase carried by bean flour and the delayed addition of salt.[3]

This artificial maturation takes place rapidly, reinforced as it is by the action of ascorbic acid and

Figure 4–3 "Punching Down" or Folding Dough.

"Punching down" a dough: this English phrase is far too brutal a term for what should be in fact a gentle folding over of the dough, stimulating fermentation and reinforcing the dough's strength. This folding should be done as indicated in the recipes, but the baker should note that an exceptionally weak flour might require an extra folding, while an exceptionally strong flour might require very gentle treatment (Figure 4–3).

---

given added impetus by the presence of an appropriate amount of yeast. Artificial maturation saves processing time by reducing the time allocated for primary fermentation, and it promotes the production of loaves with large volume. Unfortunately, because of this unnatural oxidation, a considerable loss of quality is perceptible in regards to both taste and aroma. Although it is possible to make artificially matured bread that is attractive from an exterior point of view, the crumb will be light-colored, unnaturally insipid, and often even bad tasting. Keeping quality and the overall combination of aroma and flavor will also be considerably affected.

THE INFLUENCE OF DIFFERENT BREADMAKING METHODS ON TASTE

Before beginning a discussion of this problem of breadmaking methods, it is indispensable to point out the necessity of avoiding, as much as possible, anything that might diminish the fundamental quality of dough and bread, whether at the mixing or the makeup stage. Because of the incalculable damage represented by hyperoxidation of dough (brought about by the abusive nature of intensive mixing in combination with an excess of lipoxygenase and the delayed addition of salt), I feel it is necessary to give a stern warning about these matters and to discuss seriously the taste of bread that is washed out and unnatural and that threatens to reduce the quality of French bread to the lowest common denominator, thus contributing to its eventual destruction.

Thus this chapter discusses only those breads that are produced from controlled-oxidation doughs—that is, doughs in which the degree of mixing oxidation has been controlled and maintained within reasonable limits. This controlled oxidation is not harmful to the quality of the crumb, which will be of a light cream color. When considering the amount of mechanical working of the dough to be applied, the technical requirements necessary for development of the gluten network are carefully taken into account.

The amount of mechanical working is based on the total number of revolutions achieved by the mixing vanes or arms. This must be determined for two or three types of mixers and for the mixing time in both first and second speeds (Table 4–1). Discussion in this area will be limited to what corresponds to what is known in France as "improved mixing."

Two of the types of mixers shown in Table 4–1 are well known in France: the oblique axis mixer and the spiral mixer. Two of them are well known in Brazil: the spiral mixer and the high-speed mixer.

Regarding the latter, a copy of the English Tweedy mixer, it is possible to fully develop the gluten network and correctly mix the dough by adapting to the high-speed mixer's rapidity. It has the very serious drawback of causing a strong friction temperature rise in the dough mixture, and it may be used only with very cold formula water.[4]

I should add that the total mixing times may be further reduced by at least 15% through the use of the autolysis rest period , bringing the respective totals to 13 minutes with the oblique axis mixer, 6 minutes 30 seconds with the spiral mixer, and 2 minutes 15 seconds with the high-speed mixer. These reductions will further tend to diminish the oxidation of the dough, to the benefit of good bread taste.

It must also be noted that the method of bread production certainly has a determining influence on bread taste. First of all, there is a distinction between breadmaking with a natural sponge starter and breadmaking with baker's yeast.

Table 4–1 Comparative Table for *Improved Mixing:* Various Machines

	Oblique Axis Mixers	Spiral Mixers	Artofex	20-quart Planetary	80-quart Planetary	Tweedy or Brazilian High-speed Mixers
First (slow) speed	40 RPM	100 RPM	40 RPM	100 RPM	Second speed 100 RPM	400 RPM
Mixing time	3 min	3 min	4 min	3 min	3 minutes	2.5 min
Total revolutions	120 revolutions	300 revolutions	160 revolutions	300 revolutions	300 revolutions	1000
Second speed	80 RPM	200 RPM	70 RPM	200 RPM	Third speed 180 RPM	—
Mixing time	12 min	4 min	11 min	4 min	4.5 min	
Total revolutions	960 revolutions	800 revolutions	770 revolutions	800 revolutions	810 revolutions	
Total mixing time	15 min	7 min	15 min	7 min	7.5 min	2.5 min
Total revolutions	1080 revolutions	1100 revolutions	930 revolutions	1100 revolutions	1110 revolutions	1000

Please note: The above is only a guideline. In the text, Professor Calvel lists mixing times for each recipe for oblique mixers, unless specified otherwise. Mixing times vary according to machine type, flour quality, the presence of preferments, batch size, bowl and mixing arm friction, etc. Home or retail bakers using commonly available Hobart or Kitchenaid machines should adjust mixing times as indicated for planetary mixers. Otherwise, overmixing will occur, with unfortunate results.

Breadmaking with a *Levain* or Natural Sponge Starter (Sourdough)

Production with a *levain* includes both breadmaking with a naturally fermented sponge[5] and breadmaking with a *levain de pâte*, or mixed sponge (with added baker's yeast). As previously indicated, breadmaking with a *levain* is an acid fermentation, which under certain weather conditions may give a pronounced sharp taste to the bread.[6] This very noticeable and slightly bitter character sometimes becomes excessive.

Whenever leavening is accomplished only by means of a *levain*, the bread is thickly compact in form, most often round—as in the case of miches or tourtes—or slightly elongated, as are **boulots**. Generally, they are relatively dense. Since oven spring at the beginning of baking is slower, the round form is more suitable for expansion of the loaf and allows the baker to better overcome the handicap implicit in slow oven spring. In any case, when the starter sponge and the dough are in good condition—it is better if they are not too firm—the baker may obtain attractive, well-raised breads from them, with a supple and elastic crumb structure. In addition these breads have a light, cream-colored crumb that has an agreeable aroma and taste and has the advantage of long conservation of both freshness and taste.

In addition to the natural *levain* method, which is accomplished at all stages without any purposeful addition of baker's yeast, a parallel breadmaking method has developed that involves use of a *levain* with the addition of a small amount of baker's yeast. This production method is known as *levain de pâte* breadmaking, or the mixed sponge ***levain mixte*** method.

Depending on available yeast transportation facilities, this method has been adopted nearly everywhere in France, especially during the cold season when fermentation is slower and more laborious. It allows time savings at all the different stages of fermentation. Furthermore, since the oven spring is much more pronounced, this method provides excellent conditions for the production of long breads with baked weights of 700 g or 400 g, as well as baguettes.

As far as taste is concerned, bread produced using the *levain de pâte* method shows a noticeable loss in vinegar odor, allowing greater presence of the aromas derived from the flour itself. The flavor elements contributed by the action of baker's yeast during the fermentation process are also more apparent, as are those that come from the actual baking of the dough. This type of bread is an agreeable and highly acceptable compromise for the majority of consumers, although the acetic odor is discernible and may even become excessive in summer. This method also produces breads that are characterized by good keeping qualities, as much from the perspective of freshness as from that of aroma. Of course, this is always on condition that the dough has not been the victim of abusive levels of oxidation during mixing.

It should be noted that to ensure a regular production schedule, the baker is required to make up and maintain each day the starter sponges that will be used to leaven the doughs for the next day's production. These sponges are started from the chef—a portion of dough taken from a previous day's batch after fermentation.

To maintain the viability of the culture, it is necessary to ensure that the temperature of the refrigeration chamber stays between 8 and 10°C (46.4°F to 50°F) whenever the chef is retarded for periods of 48 hours or more. At lower temperatures, part of the flora of the culture may be destroyed, and consequently the taste of bread produced from this culture may be spoiled.[7] This same consideration is true, although to a much lesser extent, for the *levain de pâte* method. Thus, bread made from a *levain* frequently no longer has the dominant, lightly acidic taste that is the basis of its originality.

Breadmaking by Leavening with Baker's Yeast

The use of baker's yeast has given rise to several methods of breadmaking. In general, three of these are used in the production of French bread: the **poolish** or Polish sponge method, the **levain-levure** or sponge and dough method, and—far and away the most widespread—the **méthode directe** or straight dough system. This last may be improved on occasion by adding a proportion of prefermented dough.

The first two methods are older. Both involve the use of fermented cultures prepared in advance of dough mixing, as with the *levain* method.

During the 1840s in France, beer yeast, or baker's yeast, began to be produced especially to meet the needs of bakers. When making dough that used yeast as the only leavening agent, bakers logically proceeded according to the same steps as they had with sourdough cultures.

Breadmaking with Polish-Style Sponge (Poolish)

The first of these baker's yeast based methods to become known in Europe appears to have been the Polish-style sponge, or *poolish*, method. The *poolish* is a relatively liquid fermented culture, leavened with baker's yeast in advance of dough mixing and prepared with only part of the flour and water. When the dough is made, the rest of the flour, water, yeast, and salt are added, and an appropriate production schedule is then followed until the baking of the bread is completed.

This method of breadmaking was first developed in Poland during the 1840s, from whence its name. It was then used in Vienna by Viennese bakers, and it was during this same period that it became known in France. The bread produced by this method became known as Vienna bread, named for those Austrian bakers who began to make it in Paris after having come from Vienna. The *poolish* method seems to have become the only breadmaking method used in France until the 1920s for the production of breads leavened solely with yeast.

On the storefronts of very old bakeries, one can still see advertising signs that enticed consumers with the words "French bread and Vienna bread." The former was bread produced from a natural starter, most often from a *levain de pâte*. The latter was bread based on baker's yeast and produced by leavening with a Polish-style sponge. This sponge had to be prepared separately for each batch, or **fournée**.

The Vienna bread produced according to the *poolish* method was much less acid to the taste than bread leavened with a levain or with *pâte fermentée*, and the acetic or vinegar odor was also absent from it. Furthermore, the fermentation of the sponge continued for 3 to 7 hours—sometimes more—before mixing, resulting in the formation of complex organic acids that enriched the aromatic content of the dough and the resulting bread. It was this enrichment from a nonacetic alcoholic fermentation that made the bread so tempting while allowing the taste and aroma characteristic of the pure wheat flour in use at that time to be fully appreciated. When the flour was hypodiastatic or low in fermentable carbohydrates, a little malt extract was added during mixing as a remedy, and the aromatic effect of this addition was significant. In addition, this bread had very good keeping qualities.

Levain-Levure, or Sponge and Dough Breadmaking

It would well to remove any ambiguity as far as vocabulary is concerned at this point. The term "sponge" has doubtless been used for this method because it refers to a fermented culture made from flour, water, and yeast; it is relatively firm in consistency, like a naturally fermented sponge, in contrast to the more liquid *poolish*. As the name also indicates in French, leavening is provided only by *levure* or baker's yeast. As in the case of the *poolish*, this sponge must be made up in advance of each batch of dough. In any case, this method has been used very little in the production of French bread, even though it certainly could be. It has been used a good deal—and still is—in the leavening of yeast-leavened sweet doughs, and more particularly in the production of English tinned bread.[8]

The sponge and dough procedure was long favored as the basic production method for **white pan bread**. This method contributed to the bread a degree of quality, taste, and shelf life that has since been lost because of the adoption of the straight dough method, which is used in England, the United States, and other Anglo-Saxon countries today. This quality loss has occurred in spite of—or perhaps because of—the presence of numerous additives used in the straight dough production method. There is a single exception, which certainly deserves to be highlighted. Japanese bakers continue to produce white tinned bread with the addition of a prefermented yeast sponge, and today make the best pan bread in the entire world.

The sponge and dough method was also used—and still is—for the production of common brioche, and especially for the Vendée-style brioche. Finally, the sponge and dough method is frequently used in the production of country-style breads, in which it produces excellent results.

The sponge and dough procedure may find a place as well in the production of traditional French breads; the results are close to those obtained with the *poolish*. As in the case of the latter, the sponge and dough method enriches the dough with organic acids. Consequently, it yields a bread that by taste is clearly related to other breads leavened with baker's yeast, without a perceptible aroma of acetic acid. The sponge and dough method improves the flavor and aroma complex produced by alcoholic fermentation and the aromas contributed by wheat flour and the baking process itself. These latter characteristics continue to develop, become more pronounced in the finished loaf, and yield excellent results in respect to both taste and shelf life.

Direct "Straight Dough" Method

This is the most recently developed breadmaking method, and today is the most widely used for bread production. In this case there is no previously prepared culture, and all of the ingredients are placed in the mixer, either prior to or during the mixing stage. As in the case of both the *poolish* and the sponge and dough methods, each batch is mixed independently of others.

It was shortly after World War I—during the 1920s—that this method of breadmaking was first conceived. At that time, the bread thus obtained was of excellent quality, and this more simple method of direct addition of leavening began to be substituted for the *poolish* method. It also began to gradually replace the *levain de pâte* method in urban areas and eventually even in rural districts. It was this high quality, which was to be maintained for the most part until 1940, that would form the basis of the reputation of French bread around the world. The artisan bakers of Paris played an outstanding role in developing and maintaining this reputation.

In this simplified makeup method, the dough is leavened with a relatively small amount of yeast, around 0.8% based on flour weight. The weak flours of that period—from 80 to 100 W on the average—made dough production a rather simple matter. At that time mixing was done with slow single-speed mixers, and sufficient mechanical working of the dough was accomplished by the alternation of mixing and rest periods—three work periods of around 5, 4, and 3 minutes and two rest periods of 5 minutes each between the mixing stages—for a total of 20 to 25 minutes.

The dough was hardly oxidized at all by this slow mixing method and remained cream-colored throughout. Natural oxidation and maturation were accomplished by a prolonged **first fermentation** or **pointage**, from 4 to 5 hours. To improve the tenacity and the elasticity of the dough, this stage included punching down the dough about two-thirds of the way through the fermentation period. The dough was weighed and molded manually at the end of the first fermentation. The **second fermentation** of around 60 minutes was rather brief by comparison with the first and produced a bread that was rather low in volume.

The length of time required for bread production by the straight dough method was about 6 to 7 hours, all inclusive. As far as taste was concerned, the bread was very flavorful. It had a taste a little less pronounced than bread made by the *poolish* method, but it was just as appetizing—perhaps even more so. This inherent lightness of taste allowed greater scope to the aromas that originated from the wheat flour, as well as from the long, slow alcoholic fermentation that took place throughout the whole of the dough.

Whoever had the privilege of biting into a hard roll, a **bâtard** loaf, a **baguette**, a **marchand de vin** loaf, or— until 1937—a **4-pound split loaf** would never forget the pleasure. (It was in 1937 that this type of large loaf had to be abandoned, as it had a high production cost, but it was priced too inexpensively.)

The secret of this quality was a dough that was not oxidized during mixing and an alcoholic fermentation that was carried out slowly and completely, producing those precious organic acids whose presence both enhanced the taste and ensured a reasonable shelf life. This explains to a great extent the substitution of the straight dough procedure for the *poolish* method and the gradual abandonment of breadmaking from a *levain de pâte*.

However, from the beginning of the World War II (1939–1940), and for the next 7 years, unsuitable flours and practical problems brought on evolutionary changes that were harmful to quality bread. The processes of production became more rapid, and from 1958–1960 onward, two-speed mixers and overmixing began to appear, along with the unfortunate consequences we still experience today. It is certain that the straight dough method bread produced under such conditions suffers greatly from overmixing and from the "washing out" of bread taste that results.

In the face of such a change, the bakery industry of today has often been found unprepared to deal with the demands of social change and of general technological progress.

Straight Dough Method with Addition of Preformented Dough

However, an effort to adapt the straight dough method can still allow bread made by this method to remain authentic and to preserve both its distinctive nature and its high quality. This acceptable variation on straight dough breadmaking includes the addition of prefermented dough taken from a previous batch during the course of daily production. While this is not by any means a new idea, it permits good and quick production of quality bread with almost no problems at all. Furthermore, this adaptation makes use of the mechanical equipment already available to today's baker with a minimum of complications.

The amount of prefermented dough to be added may vary around 15% of the weight of total formula flour, sometimes a bit more. It should have a minimum of 3 hours of fermentation after having been leavened with 1.5% yeast (based on total flour weight). The prefermented dough may be used for up to 12 to 14 hours when held at ambient temperature; when kept under refrigeration at 4 to 5°C (39.2 to 41°F), practical use for up to 18–20 hours and more is possible.

This addition allows a reduction in the first fermentation period and permits the baker to accomplish before mixing what he or she either cannot do or no longer has the time to do after mixing. It must be understood that any amount of excessive oxidation of the dough during mixing must be avoided in order to fully benefit from the organic acid enrichment and the improvements in both taste and shelf life made possible by maturation of the dough.

If all these precautions are taken, bread produced by this method will again exhibit the characteristics of straight-dough, yeast-leavened bread produced with a prolonged first fermentation, or by extension, those of the *poolish* or sponge and dough methods.

Exhibit 4–1 Base Formula for Prefermented Dough

> Flour: 100%; 580 g (20.458 oz)
> Water: 69%; 400 g (14.109 oz)
> Yeast: 1–2%; 5.8 to 11.6 g (0.20459 oz to 0.4091 oz)
> Salt: 2%; 11.6 g (0.4091 oz)
> TOTAL DOUGH WEIGHT: 1000 g or 1 kg (35.273 oz)
> Mixing: Low speed: 4 min
> Second speed (beater): 3 min
> Dough temperature from mixer: 25°C / 77°F
> Prefermented dough rest time: 3 h @ 25°C / 77°F *or*
> 16 h @ +4°C / 39°F

Prefermented Dough

Although in the past many bakers rather haphazardly kept a piece of old dough to put into the next batch, the systematic use of prefermented dough is a relatively recent development that Professor Calvel has worked hard to promote. Instead of being specially prepared for each individual batch of dough, as with the poolish and sponge and dough methods, prefermented dough may simply be "borrowed" from a batch of French bread made earlier in the day (or from a batch made the day before and kept subsequently under refrigeration).

In other words, dough taken from batches made by the sponge and dough, *poolish*, traditional, improved mixing, or prefermented dough methods (see Chapter 10) would all be equally suitable for use in this procedure. However, it should be borne in mind that taking dough from one batch and adding it to a subsequent batch cannot be done indefinitely, or even over several generations, without the danger of producing undesirable flavors. It should also be noted that the use of prefermented dough which has fermented at room temperature for many hours, especially if it comprises a large percentage of the total formula (country-style bread, Exhibit 11–1, for example) can possibly lead to a lack of residual sugars in the shaped loaves and result in poor crust coloration and other problems (see Chapter 6). By the same token, a well-fermented prefermented dough (but not exaggeratedly so) can be useful when dealing with hyperdiastatic flours (see Chapter 1). These principles apply to all yeast-based preferments.

For the purpose of calculating recipe yields, the hydration of the prefermented doughs called for in the recipes in this book has been set at 68% to 69%, which is normally suitable for North American flours. Of course, the true hydration levels will vary somewhat with no effect upon the results if the dough has the required texture.

Bakers who make products from many recipes based on prefermented dough might find themselves "borrowing" too much dough from other batches for the system to remain practical. In this instance, it is preferable to calculate the amount of prefermented dough needed, and then to produce a single large batch of prefermented dough for addition to the various product batches. When the gain in quality, speed, and organizational flexibility is taken into account, it is apparent that the small extra effort of making a batch just for use as prefermented dough is very worthwhile.

The recipe shown in Exhibit 4–1, which does not appear in the French edition, is a basic guide to the production of 1000 g (1 kg) of prefermented dough at a 69% to 70% hydration rate.

Here is an overview of the valuable possibilities offered by bread production methods that involve a fermentation culture. The following diagrams (Exhibit 4–2) will help the reader to more clearly understand the procedures involved in the various methods of bread production, all of which begin with the "improved mixing" procedure. This mixing method offers a good compromise between the requirements for adequate mechanical working of the dough, those necessary to safeguard the quality of the bread, and the procedures of the different breadmaking methods. Note the exceptions for the variations on the straight dough method, which may be considered controls, one being accomplished by conventional mixing, a second by improved mixing, and the third by high-speed mixing.

EVOLUTIONARY CHANGES IN THE DIFFERENT BREADMAKING METHODS

As has been noted, until the beginning of the 20th century the great majority of bread production was achieved by leavening with *levain* or with *levain de pâte*. It was not until the beginning of the 20th century, and especially since the 1920s, that leavening with baker's yeast gained ground and rapidly came to dominate bread production. It is important for the professional as well as for the consumer to know and understand this.

Part of a presentation that I delivered on this subject at the 1973 *Europain*[9] conference describes this change:

> ...During the early 1920s a critical mutation came onto the scene, and bit by bit it began to radically change the entire structure of bread production. This change was the generalized use of biologically cultured baker's yeast as a fermentation agent. At first this was done in the form of a small addition of baker's yeast to a natural *levain*, essentially transforming it into a *levain de pâte*. During the same period, however, baker's yeast was very rapidly gaining importance as the sole fermentation agent in breadmaking.
>
> It is certain that baker's yeast was already known and had already been used as an addition to levain for a long time in larger cities.[10] It had also been used in the second manner as well, but to a much smaller extent. Bakers commonly used it to ferment the doughs of what was then called *pain viennois* or Vienna bread. This was made from fermented yeast culture known by the name of *poolish* or Polish sponge, prepared in advance of dough mixing. This type of preferment certainly had the advantage of yielding good bread but also had the disadvantage of having to be made up several hours in advance, cluttering the production area somewhat during the fermentation period.
>
> It was then that what is termed the direct or straight dough method was developed: it required no advance culturing—water, salt, yeast, and flour were all placed together into the mixer bowl. The typical single-speed mixer in use during that period mixed the ingredients from 6 to 7 minutes, was stopped for 4 to 5 minutes, and was then restarted for 5 to 6 minutes to finish mixing. The result was a smooth dough that was left to ferment about 4 hours in the mixer bowl or in a dough trough, and that was punched down and degassed about three-quarters of the way through the fermentation period. This was then followed by dough division and rounding. The second fermentation and final forming of dough pieces followed, and around 60 minutes later, it was time to bake.
>
> This bread production method rapidly invaded Paris, then became established in the larger cities of France, and then took root more slowly in the less important towns.
>
> In 1932 and 1933, in the southwest of France (or more precisely, in the Tarn region), on the occasion of an area business fair, a large mill from the Parisian region was represented, with a baker-demonstration agent and his sacks of flour. The baking was done in a bakery in the center of town, and the sale of samples took place in a booth at the fair.
>
> The bread was nice looking, and less acid than bread made by the naturally fermented dough method, which is sometimes overly so. The texture of the bread was beautiful, the color creamy, and both smell and taste were vivid and tempting. The bread was a success, although many consumers had one criticism: it was too good and as a result they ate too much of it!
>
> That is not to imply that one cannot find good bread made by the levain de pâte or prefermented dough method: when the

Exhibit 4–2 Schematic Comparison of Baking Methods

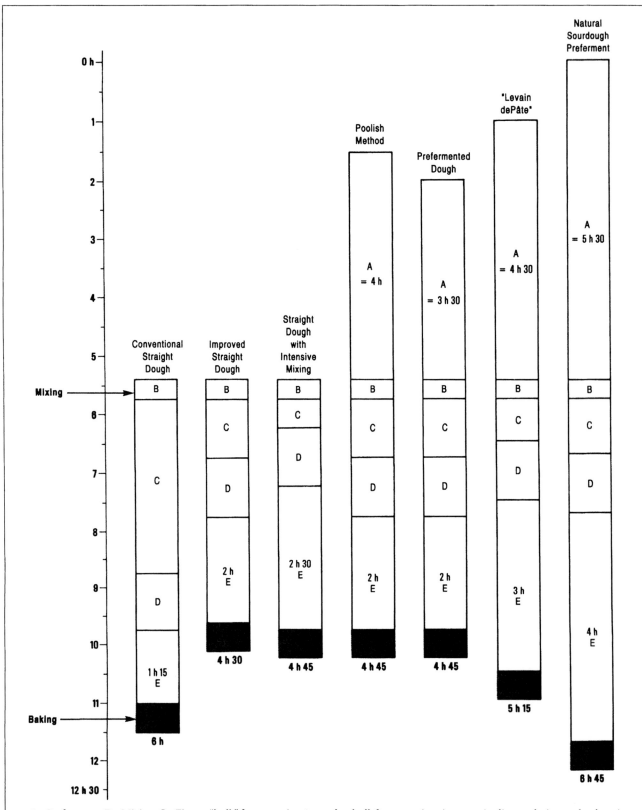

A = Preferment; B = Mixing; C = First or "bulk" fermentation (note that bulk fermentation times are in direct relation to the duration and speed of mixing); D = Dividing, weighing, forming; E = Proofing.

successive sponges have been brought to a reasonable level of maturity and are not excessively acid, bread from naturally fermented dough supports comparison with bread made by the straight dough method, but it must be admitted that bread production by the *levain de pâte* method is more uncertain and less regular in quality.

In my own case, I had the occasion to produce some good *levain* type bread during that period. In the winter it was made from a *levain de pâte*, with the addition of a little baker's yeast, while during the warm season—from May 15 through September 15—I used a natural *levain* alone, without any additional yeast.

The straight dough method, based on baker's yeast, continued to gain ground: it was simpler, required less work, and ensured a more predictable quality of production. However, except for the Parisian basin and the north and northeast parts of France, customers in rural areas continued to prefer bread made by the *levain de pâte* method. It kept better, and since it was often more dense, it gave the impression of being better able to satisfy robust appetites.

These years were truly the golden age of French bread: who then wasn't familiar with the 4-pound split loaf, the two-pound loaf, or even better, the *marchand de vin*, the baguette, the *petits pains* or hard rolls. It was especially from that period of time that the fame of French bread began to spread throughout the world. People say to me, "You were only 20 years old." That is perfectly true, and I must add that I had a very healthy appetite!

That bread, even today, has the same superior gustatory qualities as in the past. Furthermore, everyone who has had the opportunity to taste it experiences the same pleasure in eating it. The enticing qualities of this bread are universally experienced by young and old, whether at the School of Milling in Paris or elsewhere.

The secret of this quality is to be found in just a few words: during mixing, the dough is not oxidized or denatured, and it retains the authentic flavor of wheat flour itself. Furthermore, the first fermentation is initiated with a rather small dose of yeast—from 0.8% to 1% (based on flour weight)—and extends over about 4 hours or more. This lengthy period gives secondary fermentation the time to develop and to enrich the complex flavor and aroma generated by the primary fermentation carried out by fresh, industrially produced baker's yeast.[11]

As to the controversial question of the quality of bread obtained by leavening with a *levain* in comparison to bread made by fermentation with baker's yeast, I would like to affirm that both can achieve an optimal quality level. The first will possess a stronger, sharper taste, with a very slight acid nature. The second will have a finer, more delicate taste, and the aroma and taste elements derived from wheat will have a greater presence.

Thus, it becomes very difficult for me to say which of the two types I preferred whenever I think back to the 1932–1936 period when I used to make bread in the southwest of France and in Paris itself. Was it the *levain*-based that I made in the country villages of Upper Garonne, Tarne, and Aveyron, or the bread I made with baker's yeast by the straight dough method in Toulouse and Paris? I would not have been able to choose at that time, and if I asked myself the same question even today, I would have a great deal of difficulty, since I still appreciate both of them so much.

But if they are made methodically, with all the necessary care, bread from *levain* is neither better, nor more noble, than bread made from baker's yeast. Whenever the culture of starter sponges and their makeup and maintenance are properly carried out, and whenever the stages of alcoholic fermentation are appropriately and correctly observed, both types of bread merit the same respect and the same consideration by their high quality. And even though they are certainly different from one another, they remain just as appetizing today as they were yesterday.

NOTES

1. In a general sense, baking can be thought of as "fermentation management." Prefermented dough, *poolish*, and other preferment methods are ways of giving a "head start" to a new dough. This approach permits shorter bulk fermentation times without loss of quality.
2. (See Figure 4–2.) Glass mercury or alcohol thermometers should **never** be used in any food production operation for food safety and liability reasons.
3. Potassium bromate, azodicarbonomide, and oxidizing additives have much the same effect.
4. The Tweedy also has the advantage of being adaptable to mixing under vacuum, thus helping to avoid the problem of excessive oxidation.
5. Also known as a *biga* in Italian or a *sour* in English.
6. Research in the late 1960s and early 1970s by Kline and Sugihara brought to light that sourdoughs are generally composed of wild yeasts and heterofermentative lactic bacteria. Warmer temperatures encourage the development of milder lactic flavors, while cooler temperatures promote the growth of more acetic flavors. Refrigeration decimates the wild yeasts, giving more of an acetic acid character to the dough and finished baked loaves.
7. Below 8°C it is usual for wild yeasts in the culture to be destroyed, while the acetic acid bacteria will continue to thrive. See the research papers by Kline and Sugihara on *Lactobacillus sanfrancisco*.
8. Very similar to richer versions of American white pan bread.
9. *Europain*: an international trade show and exposition that is now held every 3 years in Paris.
10. The reader should remember that baker's yeast and brewer's yeast are both varieties of *S. cerevesiae*, and that the practice of using beer foam as an aid to bread fermentation has existed since the time of the early Egyptians. Malouin also mentions it in the *Art du boulanger* (1667).
11. The original paragraph was in the imperfect tense, which denotes action carried on in the past. However, since Calvel refers in the paragraph immediately preceding to the fact that the same type of bread is still made today, the present tense is more logical in a discussion of the basic methods of production.

Organic Acids

Chapter 5

- The identification of volatile organic acids and their influence on the taste of bread.

- The relationship of organic acids to mixing intensity, dough oxidation level, and bread production method.

THE IDENTIFICATION OF VOLATILE ORGANIC ACIDS AND THEIR INFLUENCE ON THE TASTE OF BREAD

There is certainly a difference between bread from a natural *levain* or starter sponge and that produced from a *levain de pâte* or prefermented dough.[1] The first method produces relatively dense bread with a predominately acetic acidity, while bread made from a *levain de pâte* is better developed, is higher in volume, and has an acidity that is still distinctly present but much more discreet and complex.

Also, the several previously described baker's yeast leavening methods each have their own characteristics. Provided that the dough has not been excessively oxidized, bread produced by the straight dough method has a moderate flavor that is still agreeable and appetizing. This is especially so if the percentage of yeast used is relatively high and the first fermentation period rather brief. When a prefermented yeast culture is added, as in the cases of the *poolish, levain-levure* (sponge and dough), or *levain de pâte* methods, the bread generally has a more pronounced flavor than that made by the straight dough method. Even though this stronger taste is somewhat different, it is certainly not more pleasant, especially whenever the latter is made without excessive dough oxidation during mixing, and when the amount of yeast used is proportionally comparable overall.

Some interesting investigations have been conducted on this matter by a team from the National Institute for Agronomic Research (INRA) in France, including researchers D. Richard-Molard, M.C. Nago, and R. Draperon. These studies were concerned with the nature of the volatile acids produced by three methods of leavening commonly used in breadmaking—straight dough, *poolish*, and *levain*—and involving production by either conventional or intensified mixing. From the beginning of the culture fermentation or from their incorporation at the beginning of mixing, the three methods had total fermentation times of

- 4 hours for the straight dough method (using 2% yeast based on flour)
- 8 hours 30 minutes for the *poolish* method (using 0.7% yeast based on flour)
- 15 hours for the natural *levain*

There was also a small amount of yeast used in the case of the natural *levain*, but the researchers did not specify the amount.

The analyses giving the final results discussed by the authors were of two types:

1. an examination of the volatile constituents of the bread crumb, especially the organic acids as shown by chromatographic analysis
2. an investigation of the aromatic characteristics of the constituents of the crumb, obtained by olfactory analysis. This was accomplished by evaluating the chromatographic profile recorded in column rank at the end of the study.

The first of these two analyses allowed the researchers to very incompletely identify the nature and relative proportions of the different volatile products that occur in bread crumb. The second permitted them to conclude that none of the individual components of the aroma of bread crumb is identifiable as bread aroma in itself, but that it is instead the combination of a number of different volatile products that the crumb contains. From the results, the authors were able to compare the breads produced from *levain* with those leavened with baker's yeast, and for the latter, to determine the differences between bread made by the *poolish* method and that produced by the straight dough method.

They noted on this account: "Remember that the methods of breadmaking which include a prefermentation stage, like the *poolish* or *levain* methods of production, have the reputation of leading to more flavorful products with a more agreeable aroma than those obtained from straight dough breadmaking." (C.N.E.R.N.A. 1977)

"It is generally admitted that these superior qualities result from long periods of fermentation that permit the processes of secondary fermentation to occur with greater intensity. It is these secondary fermentation processes which are normally generators of numerous volatile composites." (Bourdet, 1977) Tables 5–1 and 5–2 give the average levels of volatile organic compounds that are found in the crumb portion of bread.[2] Table 5–1 refers to conventional mixing, while Table 5–2 is for intensive mixing.

These results led them to remark that "whatever the type of bread, the major component is acetic acid, followed distantly by propionic acids, isobutyric, butyric, isovaleric, valeric and capriotic acids. The authors have identified other acids, but from the acid of heptane on, their volatility and water solubility become so weak that there is little probability of any involvement as an aroma component."

They state that even though the characteristics of the three chromatographic profiles of the volatile acid fractions of the three bread crumb types under study have the same qualitative composition, under conditions of actual consumption significant quantitative differences may be noted, particularly in regard to acetic acid (see Table 5–1).

Average Levels of Volatile Organic Acids in Bread Crumb

"In comparison to the straight dough method, the *poolish* and *levain* methods produce acetic acid levels that are respectively two and twenty times greater. The other acids are only present in very small quantities, and show much smaller differences from one type of crumb to another. However, it is apparent that the level of isobutyric acid is around four times stronger in the crumb of bread fermented by baker's yeast than by bread made by the *levain* method. These figures support the thesis that the organic acids formed through fermentation—especially acetic acid—play an important role not only in terms of flavor, but also as an aroma component. This is why we have attempted to precisely determine the development of these acids, as a function of the principal parameters of bread production."

That led them to examine the influence that mixing, the amount of yeast used, and the length of fermentation might have on the taste and aroma of bread.

THE RELATIONSHIP OF ORGANIC ACIDS TO MIXING INTENSITY, DOUGH OXIDATION LEVEL, AND BREAD PRODUCTION METHOD

Average Levels of Volatile Organic Acids in Bread Crumb

The intensified mixing method brings about a modification of the acid fraction, as recorded in Table 5–2. The most significant differences are to be found in the cases of isobutyric and isovaleric acids. The levels of these substances found in the crumb of bread produced by the straight dough method with intensified mixing are around 3.5 times greater than the level produced through slow or conventional mixing.

Under the dual effects of intensive mixing and the presence of bean flour in the dough, the *poolish* method also brings about an increase in the levels of iso-acids. However, this increase is far below that which is seen when intensive mixing is used on a *levain de pâte*. Considering the sharply disagreeable olfactory characters of these acids, the authors of the study believe that this interesting phenomenon may help to explain why the effects of intensive mixing

Table 5–1 Conventional Mixing with 0.7% Bean flour

	Straight Dough (2% Yeast Level)	Polish Style Sponge (0.7% Yeast Level)	Naturally Leavened Starter Sponge
Acetic acid	53	105	970
Propionic acid	0.50	0.54	0.68
Isobutyric acid	1.30	1.12	0.32
Butyric acid	0.18	0.29	0.16
Isovaleric acid	0.51	0.48	0.43
Valeric acid	0.13	0.29	0.16
Capriotic acid	0.84	1.06	0.58

Quantities expressed in parts per million in relation to the crumb.

Table 5–2 Intensified Mixing with 0.7% Bean Flour

	Straight Dough (2% Yeast Level)	Polish Style Sponge (0.7% Yeast Level)	Naturally Leavened Starter Sponge
Acetic acid	55	123	1093
Propionic acid	0.70	0.70	0.90
Isobutyric acid	4.70	1.44	0.77
Butyric acid	0.26	0.18	0.38
Isovaleric acid	1.80	0.56	0.55
Valeric acid	0.16	0.16	0.24
Capriotic acid	1.10	1.00	0.80

Quantities expressed in parts per million in relation to the crumb.

may be less apparent when producing bread based on a *poolish*.

The amount of yeast used exercises some influence as well, since it also affects the volatile organic acid content of the crumb. The research performed allows the observation that

1. when the 2% yeast level, which is customarily used in straight dough breadmaking, is increased to 2.2%—as is done by some bakers—intensive mixing causes an increase of nearly 50% in the levels of isobutyric and isovaleric acids in the crumb, while the other acids—acetic acid in particular—do not change at all.
2. with the *poolish* method, where the yeast level is normally less (but *not* at the level used in this particular N.D.L.R. study), the use of more leavening leads to an increase in not only the iso-acids content of the bread but also the content of certain linear chain acids. Thus, the content of acetic acid increases in relation to the percentage of yeast used and reaches a maximum value of around 200 parts per million from the use of 1.6% yeast.

The influence of total fermentation time, which is measured from the mixing of the dough until the unbaked loaves are placed in the oven, is equally evident. On this subject the authors indicated that "these analyses refer only to bread crumb produced by the straight dough method, involving the use of intensified mixing and bean flour, with a yeast content level of 2%."

They continue, "As noted previously, the only significant variations have to do with acetic acid and the isobutyric and isovaleric acids. The content of acetic acid increased in a linear fashion during the first six hours to become stabilized at a maximum level on the order of 75 to 80 parts per million. On the other hand, the content of iso-acids increases in a regular fashion for up to at least ten hours of fermentation, but the experiments were not continued beyond that point."

After examining these results carefully, I would like to redirect the interpretation of these conclusions very slightly, without placing the general results in doubt in any way.

1. I have some reservations concerning the quantities of baker's yeast that were used in both the straight dough and *poolish* methods to obtain the bread crumb on which the analyses were performed. In the case of the *poolish*, which is a prefermented yeast culture made up before mixing the dough, it is true that the amount of yeast used as a leavening agent is normally less than the amount of yeast used with the straight dough method. However, if the amount of yeast for a *poolish* is limited to 0.7%, it could reasonably be used proportionally at a level of 1% to 1.2% for a comparative straight dough, and not at 2% as was done. If the 2% figure were to be retained in accordance with commonly accepted usage, it would be necessary to set the amount of yeast used in the *poolish* at a minimum of 1.4%.
2. I am also concerned regarding the total fermentation time, which was *reduced* in the case of the straight dough process, while it was greatly *prolonged* in the *poolish* method. There is no doubt that important variances occurred as a result of this, especially insofar as the production of acetic and (especially) iso-acids is concerned.
3. It appears to me that the preceding remarks are given further support by the authors' statement that by increasing the amount of yeast used only from 2% to 2.2%, there was a direct increase in the content of isobutyric and isovaleric acids on the order of 50%.
4. As for the stabilization of acetic acid production in a yeast-leavened straight dough processed by intensive mixing, if the acetic acid level was able to reach a maximum of 75 to 80 ppm at the end of the first 6 hours, that could only represent a relatively early developmental stage. Whenever dough is kept under ambient conditions, at a temperature of around 27°C (81°F) for 24 hours, an acetic acid odor usually does not take long to appear.

The traditional *poolish* method as described has yeast levels that range from very low to low (0.7% to 1.2%) and relatively long fermentation periods. If it is admitted that *poolish* allows the baker to offer to the consumer bread that has a savor and aroma that seem to be superior to that made by the common straight dough method, it is likely that the use of proportionally comparable yeast levels for straight dough in combination with a lengthened fermentation period would result in different data than those that were recorded. There is no doubt that the results would be similar, and perhaps even comparable, to the data for *poolish method* products.

That is what the work undertaken at the same period by B. Launay of the National Superior School of the Agricultural and Food Industries (ENSIAA) seems to confirm, as does that carried out by M. Hourne of the Quality Control and Fraud Repression Service in "The Flavor of Bread Crumb and the Contribution of Sensory Analysis in the Presentation of Evidence on Flavor Differences Linked to Bread Production Technique."

These investigations were conducted with the aid of an unbiased taste panel. Among other matters, the project involved the examination of bread produced by three production methods: one straight dough trial, which was leavened with bakers' yeast at a 4% level based on flour, and two variations of the *poolish* method, one leavened with 2% flour and the other with 4%. The test baking was carried out in two different production shops.

After an evaluation of the bread samples, the taste panel found that

- in the first shop, the breads made in one shop by the straight dough method with 4% yeast, compared with those made by the *poolish* method with 2% yeast as well as with 4% yeast, were significantly different, but the panel did not award any clear preference to any of the variations (total of 24 responses); and
- in the second shop, all of the breads produced were perceived as being identical (total of 22 responses).

Thus, for the first shop, the taste panel noted that the different breadmaking methods and yeast levels might have an influence on bread taste, but without showing any consumer preference. In the case of the second shop, the reader can gain some idea of possible differences that might occur between *subjective* analysis by the taster and the *objective* analysis obtained through chromatography. Since the panel examined the sensory phenomenon of taste, this result is not surprising.

There is certainly reason to be cautious about the judgments of the taste panel in the ENSIAA study, as noted above. In the case of the previously discussed study of the INRA researchers, the effects of the noted variables and conditions under which they occurred leave no room for uncertainty. The differences in the yeast use levels between the straight dough and *poolish* methods (more than 130%) and the total duration of fermentation—which ranged from a single fermentation period for one method to one that was double in length for the other—cannot have been without consequences for the accuracy of the study!

If we were to return to the 1920–1930 period during which the straight dough method replaced to a great extent the practice of breadmaking with a *levain*, and almost completely replaced the *poolish* method, we would find bread that resulted from a slow conventional mixing followed by a fermentation period that extended in total duration from 6 hours to 6 hours 30 minutes. This bread was in no way inferior in terms of aroma and taste to bread made by the *poolish* method—rather the contrary. This shorter fermentation time may explain in large part the abandonment of the latter method in favor of straight dough processing.

It may be useful to again point out that among those breadmaking methods that include a fermented culture prepared before dough makeup, it is possible to successfully use either sponge and dough or *pâte fermentée* (prefermented dough) on an equal basis, provided that the *pâte fermentée* is not the result of intensified mixing. This prefermented dough should be taken from one of the preceding mixed batches, having undergone a "rise" of 3 to 5 hours or more under ambient conditions. This may be extended to 18–24 hours under a storage temperature of 4 to 5°C (39 to 41°F).

With this procedure it is possible to achieve results that are very similar to those obtained from the use of *poolish*, and this can be improved even more, provided that mixing intensity and time are kept at reasonable levels and that the baker knows how to avoid using bean flour.[3]

The conclusions of these authors are interesting and certainly worthy of being known, in spite of their outright condemnation of straight dough at the end of their work. Remember that theirs was a straight dough with a relatively high yeast content and was thus penalized by a rapid fermentation. It was also the victim of intensified mixing, of the lipoxygenase associated with it, and of the "washing out" of the dough that results. For these reasons, their condemnation appears to be both unjust and a bit excessive.

Here are the essential parts of their conclusion:

> The analyses which have been carried out within the restrictions of this joint project have permitted the identification of a certain number of points which we will restate by way of conclusion.
>
> The qualitative and semi-qualitative composition of the volatile fraction of bread crumb has been established in part. Even though numerous constituents which are present in trace form have not been identified with certainty up to this point, their olfactive characteristics are known, and their contribution to the formation of the aroma of the crumb may be estimated.

In any case, whatever the method of bread production utilized, and although the taste panel member is able to distinguish the corresponding products well, the volatile fraction of the crumb undergoes little variation, except concerning the acetic acid levels and, to a lesser extent, isobutyric and isovaleric acids. Thus, it seems that only these minor differences in the volatile acid levels of the crumb need to be considered in order to explain the differences observed between the different types of bread crumb analyzed.

This has led us to formulate the following hypothesis: acetic acid may serve as a carrier for bread crumb aroma, sensitizing the taster to the other constituents of the aroma. This effect is directly linked to the quantities of acetic acid present in the crumb.

Isobutyric and isovaleric acids are present only in very minimal quantities, on the order of several milligrams per kilogram of bread crumb, but they display important relative variations from one breadmaking procedure to another. Taking into account their marked olfactive qualities, which are very disagreeable in strong concentrations, their contribution to the aroma of bread may be significant.

In spite of numerous points which remain to be examined further, the results compiled in the course of this joint project confirm overall the essential role of fermentation in the formation of the characteristic aroma of the crumb of French bread. This is contrary to the conclusions of a good number of foreign works, according to which the fermentation period of the dough may be shortened and even eliminated without problems.

Our results indicate that an excessive reduction in fermentation time would tend to bring on a reduction, or even a perversion of the taste. On the other hand, prefermentation by bakers' yeast or with a *levain* allows the baker to obtain more flavorful products, with a more pronounced aroma than bread prepared by the straight dough method, especially when mixing is intensified.

I would personally add as well that this is true whenever a good part of the first fermentation magically "disappears," as often happens currently, but there is still more. Being from Upper Languedoc, a region where breadmaking based on *levain de pâte* has continued and is still practiced today, I can testify concerning the influence of acetic acid on the taste of bread, even though this breadmaking method is one that has suffered the most from intensified mixing. The intensity of the acetic taste in bread has increased, while at the same time culture fermentation has lost its importance. It is especially the taste of vinegar, without a trace of any other flavor, that has become so strong that it is foul-smelling and disagreeable.

The taste of bread made by the *levain de pâte* method, in combination with conventional or even improved mixing, was something else altogether—and is to this day. It has a complex flavor that is very slightly bittersweet, associated with other aromas that make it a fragrant, agreeable, and appetizing bread.

Thus, we must preserve the use of fermentation for French bread, and we should always be aware that the method of dough leavening has an important relationship to taste, since in many cases it determines fermentation time. However, the taste is influenced to an even greater degree by the intensification of mixing and consequent overoxidation of the dough, and it may be greatly damaged by these factors.

We have also seen that intensive mixing, aided by the action of the lipoxygenase carried by bean flour, brings on change in the product, with an increase in the isobutyric and isovaleric acid content of the crumb. These substances are carriers of unpleasant odor components, and this increase was shown to be greater for the straight dough method than for the *poolish* method. It is true that this particular *poolish* was proportionally rather low in yeast, while with the *levain* method yeast was hardly even perceptible.

We have also seen that the results for the straight dough and *poolish* methods were achieved with proportionally very different yeast use levels and fermentation times. In the case of the straight dough method, which was carried out with the use of both intensive mixing and bean flour, a slight increase in the yeast level brought on a considerable rise in the iso-acid content of the dough, tending to diminish the flavor of bread produced by this leavening method. Consequently, bread made by the straight dough method is doubly damaged, both by the destruction of carotenoid pigments and oxidation of the polyunsaturated fatty acids of the flour and by the production of an abnormally high level of foul-smelling iso-acids. Even bread leavened with a *poolish* but intensively mixed could hardly undergo any less damage.

The damage is thus not due to the leavening method, but to the manner in which production takes place. With slow mixing, the amount of iso-acids found in the crumb of breads produced both by straight dough and from a *poolish* is somewhat comparable, in spite of the distortion from differing yeast use levels and variations in fermentation time. Where the observed differences begin to become more apparent, they are due to the production manner, i.e., intensive mixing, with the inclusion of bean flour, along with dissimilar and noncorrelative yeast usage levels and fermentation periods (Table 5–2).

This explanation corroborates in part the results noted for the taste panel. With higher yeast use levels (near or identical to those used for straight doughs) and intensified mixing, one of the taste panel trials confused the flavor of loaves made from a *poolish* with those made by the straight dough method. It should be pointed out that sensory analysis, regardless of the fact that it is inherently interesting, does not coincide exactly with either recorded variables or the actual deterioration of the product experienced at consumption.

I would like to add to these observations that in spite of—or because of—the presence of acetic acid in the crumb, the damage to the flavor of bread leavened by the *levain de pâte* system in combination with bean flour and intensive mixing is considerable.

Leavening methods, fermentation, and mixing oxidation should all be examined in order to avoid this deterioration of quality. The question is asked: How might we find the best means to neutralize or to avoid these injurious factors—I should even say, these assaults—which today so greatly affect this aspect of quality?

I have thought about this for a long time. In 1982, at a seminar organized in Brazil, I was invited to participate in a study of problems posed by professional education in baking. Upon being asked what constituted good bread, I had to reply in the face of the silence of my Brazilian colleagues. In substance, I said this: "As far as I am concerned, whatever the country and whatever the type of bread, good bread is that which is produced in accord with the nature of things."

In a country where powerful additives—strongly prejudicial to the distinctive nature and the quality of bread—are in common use, I was surprised by the warmly unanimous acceptance with which the audience, both professionals and consumers alike, greeted this definition. My purpose in recalling this instance is to clarify my approach and to allow the reader to understand my interest in technical adaptations that seem indispensable to preserve and to ensure the quality and originality of French bread and of bread in general.

In conclusion, it is vitally important to avoid overmixing and the accompanying bleaching and tendency toward artificial maturation of dough. It is equally necessary to allow fermentation to play its very own fundamental and varied role. We must know how to adapt, to organize, and thus to be able on a production level to make bread *well and quickly*. This is a highly practical approach to bread production, and all aspects of it are certainly within the realm of possibility.

NOTES

1. As noted previously in Chapter 4, baker's yeast has a considerable role in dough leavening effect with the *levain de pâte* method.
2. The flour included a standard percentage of bean flour, as commonly done in France.
3. This is more rarely a problem for professional bakers in North America. However, it is common for millers to add soybean flour to high-protein wheat flour intended for the industrial production of white pan bread, and soybean flour may often be found in "all-purpose" flour for home baking. The baker must bear in mind the final product when discussing flour specifications with the miller or distributor. Bleaching and oxidizing additives are much more real concerns in English-speaking countries.

Dough Maturation and Development

Chapter 6

- The influence of dough maturation level.
- The effects of changes in pH and residual sugar levels.
- The effects of loaf molding.
- The effect of type and degree of *pâton* development.
- The effects of freezing unbaked and parbaked loaves.

THE INFLUENCE OF DOUGH MATURATION LEVEL

We have already seen that maturation affects the physical properties of dough. Barring some type of unforeseen problem, the degree of cohesiveness *increases* and dough extensibility *decreases* while dough maturation progresses. When forming or molding of the unbaked loaves is carried out (after dough division and the rest period), dough maturation should arrive at a certain equilibrium between its opposing qualities of extensibility and cohesiveness. This equilibrium includes

- a degree of cohesiveness that allows the dough pieces or *pâtons* to rise symmetrically, to "proof" to a rounded form without tearing, and to reach an appropriate volume during baking. The loaves must have good resistance to deformation in order to undergo **scarification** (razor cutting of the crust) and oven loading without significant damage.
- a degree of extensibility that allows the *pâtons* to be stretched or lengthened without problems and without tearing during molding or forming. Extensibility also gives good proofing properties during final fermentation and provides good gas retention during and after oven loading. This ability to retain fermentation gases allows the loaves to reach a good state of development and results in well-raised, light, and voluminous loaves after baking (Figure 6–1).

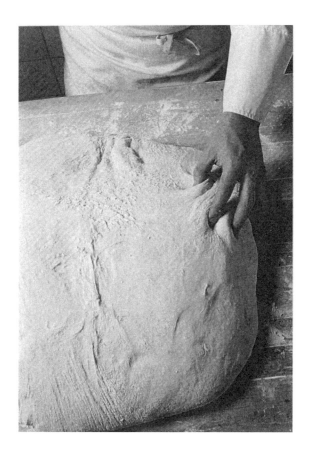

Figure 6–1 Checking Dough for Adequate Fermentation.

───────■───────

The knowledge of when a dough is sufficiently fermented requires experience and "feel."

A lack of maturation may sooner or later become apparent as a lack of dough strength, and may result in a relaxation or weakening of the *pâtons* during the course of the second fermentation. They usually proof "flat" and have a tendency to stick and to become deformed. After baking, they yield flat loaves that suffer from poor volume and have a barely acceptable appearance. The crumb structure is sometimes slightly coarse, lacks suppleness and elasticity, and very often has a marbled or irregularly colored appearance.

At the opposite end of the scale, excessive maturation—or to the baker, excessive dough strength—becomes apparent during the second fermentation as *pâtons* that proof into rounded forms, prone to crusting or drying out, and that sometimes exhibit surface cracks or tears. The loaves are very rounded in the oven, and oven spring is generally slow to develop. The incisions or blade cuts on the *pâtons* open up badly or not at all, and the crust color appears dull. Here again the loaves are barely passable or mediocre in appearance, and in general the loaf volume ranges from slightly to markedly below average. The crumb structure is usually homogeneous and evenly tinted, but this characteristic does not compensate for poor overall results.

The physical deficiencies of the loaves in both cases are evidence of a marked degradation of their gustatory appeal, like fruit that is either unripe or too ripe.

The "mouthfeel," or sensory characteristics in the mouth, also suffers in both cases. With improperly matured dough, the crumb of bread is coarse and disagreeable, while with overly matured *pâtons*, the crust is deficient. Thus, the lack of or excess of maturation of both the unformed dough and the formed, unbaked loaves results in great quality deficiencies.

THE EFFECTS OF CHANGES IN pH AND RESIDUAL SUGAR LEVELS

Progressive changes in the pH of the dough and in the level of residual sugars are consequences of increasing dough maturation and continuing fermentation. As maturation gradually progresses and as fermentation is prolonged, the dough becomes richer in organic acids, and this increase becomes evident as a lowering of its pH (Figure 6–2). These changes affect the physical properties of the dough and cause an appreciable increase in the aroma of both the dough and the resulting bread while improving its keeping qualities.

While progressive change in the dough leavened with a natural sourdough or *levain* or with a *levain de pâte* is relatively rapid, it appears to occur more

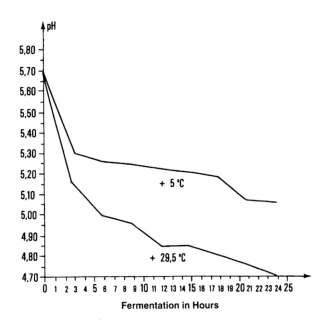

Figure 6–2 Changes in Dough pH.

slowly when the dough is leavened with baker's yeast. The curve below results from a series of pH observations carried out for 24 hours on two samples of the same dough, leavened with 2% baker's yeast. One test sample was kept at 29°C (about 86°F), the other at 5°C (approximately 41°F). The recorded results are significant.

The increase in pH is relatively rapid during the first 3 hours, but has a tendency to slow down after that time. After 15 hours, dough maintained at 29°C (about 86°F) may exhibit some changes in physical characteristics, possibly including a notable degradation of its plastic properties. Because of the presence of salt in the dough, however, the production of acetic acid is only slightly apparent. When the dough is leavened with an unsalted, prefermented yeast culture, either a sponge and dough or a *poolish*, it seems that the acetic acid or vinegar odor appears a little more rapidly, although it is still hardly perceptible.

On the other hand, dough leavened with a *levain*—a natural starter with a flora in a good state of equilibrium—has a pH that shows a continual decline. Considering the predominately acetic organic acids produced by this type of fermentation and the aromas that result from them, such a dough exhibits excessive acidity after a shorter period. Among other considerations, it is important to avoid an excessive lowering of pH. This is why the baker must build up the fermented culture with one or more "freshenings" before achieving a suitable starter with which to leaven the first dough.

Although the results of these evaluations of dough pH are influenced by the method of leavening used, and are different from one another, in the final analysis the pH is related to the level of residual sugars present in the *pâtons* just before baking. These residual sugars are the remainder of those that fed dough fermentation, and they fulfill important functions during the baking process. The level at which they are present plays an important role in determining the extent of oven spring during the first moments of baking. They also contribute to the Maillard reaction and caramelization phenomenon that produce crust coloration.

Generally a below-average pH coincides with a lack of residual sugars, which translates into a deficiency in oven spring. This problem is evidenced by a slight decrease in loaf volume and in the end by a lack of crust coloration accompanied by excessive crust thickness. The bread will also exhibit a significant lack of aroma, the crust will have less taste, and the crumb will be slightly less flavorful. In addition, the keeping quality will be reduced.

Whether doughs are leavened with a *poolish*, by *levain-levure* (sponge and dough), or with *pâte fermentée*, whenever they undergo an excess of maturation or fermentation it is a good practice to remedy in advance the resulting lack of residual sugars. To do this, add from 0.1% to 0.2% malt extract during mixing—or on very rare occasions from 0.3% to 0.5% sucrose—to reestablish the proper sugar balance.

The opposite situation may also occur, although much more rarely. The baker may sometimes find that the *pâtons* have excessive residual sugars. This is evidenced by excessive Maillard reaction and caramelization, which appears as an abnormal reddening of the crust, a serious decline in taste and aroma, and, especially, poor eating characteristics of the crust. It is difficult to correct this phenomenon, which is often due to hyperdiastatic flours. When the chosen breadmaking method includes the use of a prefermented culture, however, it is possible to succeed.[1] An excess of residual sugars may also result from an abnormally short first fermentation, or from a lack of dough maturation of the dough, and is then much more easily corrected.

Whatever the cause, the presence of an appropriate amount of residual sugars in the *pâtons* at the time of baking is extremely important: it ensures an active oven spring, assists in dough development, and helps the loaves to reach a normal volume. It is just as valuable, if not more so, in the development of optimal crust color, insofar as it contributes to exterior appearance as well as to aroma and flavor.

The most suitable color for the crust of French bread is a yellowish gold that borders on orange at its upper level. Provided that it is baked at an appropriate temperature, it will have the proper thickness. During succeeding hours it will be as crisp as could be desired and will be blessed with the best flavor and aroma and the greatest gustatory appeal.

All things being equal, the crust that results from a lack of residual sugars will be pale in color, will suffer from limited Maillard and caramelization reactions, and will have a tendency to be thicker and with a less pronounced taste than normal. Conversely, a crust that results from an excessive level of residual sugars at the baking will have a bothersome tendency to excessive reddening. It will be too thin, will become soft very rapidly after baking, will often become singed, and will have a disagreeable flavor combined with an overly soft texture in the mouth when consumed.

The degree of crust coloration, which is certainly influenced by the baking temperature of the oven, is also directly related to the residual sugar level of the *pâtons* at the time of baking. The baker must ensure that everything works together to obtain an optimal crust coloration. This is just as true of the diastatic equilibrium of the flour as of the degree of dough maturation, the loaf molding operations, and the temperature of the oven. All of these factors contribute to producing the taste, the crispness, and the flavor of the loaf, and all of them are of the most vital importance to the quality of bread and its taste appeal.

THE EFFECTS OF LOAF MOLDING

Forming or molding the *pâtons* can have a significant effect on crumb taste, and, by extension, on that of the loaf as a whole. In comparison to manual forming, mechanical molding produces notable differences in crumb structure and exercises a lengthening effect on gas cell configuration. This inevitably has some effect on the taste of the bread. Over the past several decades, mechanical forming has almost entirely replaced manual forming in the makeup of baguettes and long loaves, and to a great extent in the production of elongated specialty breads.

In either case, forming includes practically the same stages:

- By hand
  1. degassing (expulsion of carbon dioxide gas present in the dough) and reduction of the gas cell structure, resulting in the flattening and leveling of the dough piece;

2. folding and *serrage* (radial compression) of the dough piece into itself, in order to form a relatively short, cylindrical *pâton*; and finally
3. the lengthening of the dough piece for baguettes and elongated breads by a back-and-forth rolling movement of the dough cylinder on a flat surface under hand pressure.

This forming procedure is more gradual, less stressful, and generally less brutal than mechanical molding. The internal gas cell structure of the dough piece still remains in a passably irregular state, even though it is markedly reduced. When the bread has been baked, the structure of the crumb of the bread is relatively irregular and contains large gas cells. It has a better structure and is more resistant to pressure, while remaining supple and elastic. *This is the distinctive and original crumb structure of French bread.* It is pleasant to chew, which enhances its flavor.

- By mechanical molding
  1. The dough is flattened by lamination between two metallic cylinders during the first stage, and degassing is much more complete;
  2. The flat dough "pancake" that has just been passed through the laminator, is rolled onto itself. This mechanically rolled *pâton* is much tighter than a manually molded one;
  3. The dough is lengthened. It is much more tightly compressed, and the gas cell structure is finer and much more homogeneous.

Figure 6–3 Dough Divider. Courtesy of F.B.M. Baking Machines, Inc., Cranbury, New Jersey.

Modern dough dividing machinery (Figure 6–3) handles the dough much more gently than older models, but weighing and dividing by hand (Figure 6–4) is more gentle still than machine division.

One way of defining "artisan" and "artisanal" might be to say that there are hand skills used in certain trades that cannot be learned except through the guidance and example of an expert. There are many bread-shaping methods, but the basic principles remain the same.

Professor Calvel has written elsewhere that bakers must possess "hands of steel in velvet gloves," and indeed it would seem to be working at cross-purposes to endeavor to shape a piece of dough into a tight ball or a *baguette* shape and yet retain its open and irregular cell structure.

The real key to success lies in using a dough that has fermented sufficiently to acquire the structure necessary to retain fermentation gases during shaping, and in choosing a shaping method that produces a symmetrical loaf while causing minimal damage to the interior structure.

Throughout the process, it should be borne in mind that handling and shaping dough pieces makes them "tense"—a bit like a flexed muscle. If a dough is found to be difficult to shape, it is always better to allow it to rest and "relax" for a time, rather than to risk tearing or other defects.

It is also important that proper room temperatures are maintained and that draughts be avoided at all cost, both during dough "rest" periods and during shaping or forming.

Figure 6–4 Weighing and Dividing by Hand.

After baking, the crumb structure is made up of very regularly shaped, smaller gas cells. When the bread is very high in volume, it is soft and yielding, and the resulting crumb is extremely smooth and supple. These qualities are less agreeable and the overall mouthfeel is much less pleasant. Thus, this great departure from the very nature of the traditional crumb structure tends to diminish the flavor and lessen the quality of bread made by mechanical molding methods.

Another very important factor also has a bearing on this matter: the degree of fermentation or dough maturation at the time the *pâtons* are formed. When the dough has reached an advanced state of fermentation and alcoholic fermentation has done its work, the dough has more gas cells, is more elastic, and retains its shape better through the molding procedure. At this stage it is more difficult to degas the dough, and as a result the crumb structure is more irregular, better aerated, and composed of a greater number of gas cells.

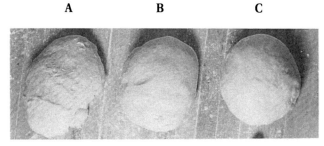

Figure 6–6 Weighing and Preshaping by Hand. (A) Weighed; (B) Preshaped; (C) Shaped Loaf (following rest).

As they are weighed (Figure 6–6A), the dough pieces are given a cursory preshaping, which basically consists of folding the edges of the dough piece underneath to form a smooth and seamless pillow shape. This is called *boulage* in French, and is the origin of the word *boulanger* (= baker). This is a gentle process, except in the case of doughs that lack strength and would benefit from firmer treatment. Preshaped loaves (Figure 6–6B) should rest for approximately 20 minutes, though the time varies according to the recipe and dough type. Very strong doughs should rest for a longer period. The dough pieces must be covered and protected from draughts. Note that baguettes and other long loaves may be preshaped into slightly oblong pieces to make it easier to do final shaping.

Round loaves are shaped by rolling the loaves along the circumference of the bottom surface, causing the lower walls of the dough piece to be absorbed into the bottom. Round loaves must be smooth and seamless and form a round ball rather than a flattened one (Figure 6–6C).

Long loaves are shaped in a stepwise fashion (Figure 6–7): (1) the loaf is turned rounded side down—seam up—onto a lightly floured surface, then flattened with the palm using a firm patting motion. (2) The outer edge is then folded to the middle of the dough piece, and the layered section is firmly patted with the palm. This is repeated a total of three or four times. (3) During the last repetition, the extended thumb and heel of the hand are used to form a "gully" or trough the length of the loaf. (4) The loaf is then turned so that the baker can use the heel of the hand to seal the seam of the loaf with a tapping motion. The result should be a resilient, bouncy cylinder with a round—rather than flattened—cross section. The shaped loaves are allowed to rise right side up (à clair) for baguettes, or upside down (à gris) for certain country loaves. In either case, it is important that at time of baking the seams of both long and round loaves always be at the bottom of the loaf.

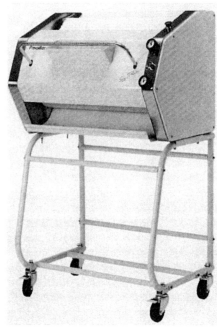

Figure 6–5 Shaping Machine. Courtesy of Pavailler, Inc., Dorval, Canada.

No shaping machine (Figure 6–5) yet possesses the firm gentleness of hand shaping. Setting the machine to a looser adjustment and lengthening the loaves by hand can improve the crumb texture, leading to a better appearance and more open cell structure. It is advisable to avoid the common French practice of shaping baguettes with exaggerated pointed ends that tend to burn and dry out during baking.

Although manual forming is relatively practicable at this stage of dough development, it is more laborious and often very difficult when it must be carried out mechanically. The mechanical method is more suitable for an extensible dough, and to use with "relaxed" dough pieces that are produced with a less advanced degree of fermentation or proofing—in fact, mechanical molding requires it. When dough is in this state, the volume of the *pâtons* is reduced less by degassing, dough forming is easier, and the tightly molded forms are better maintained. The resulting gas cell structure is more homogeneous, with smaller

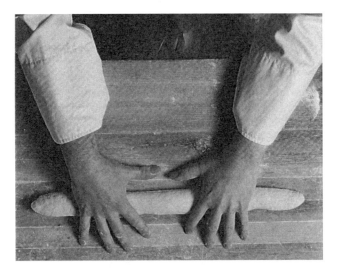

Figure 6–7 Shaping Long Loaves.

Figure 6–8 Checking for Sufficient Proof Level.

The classic method of setting baguettes and other long loaves to rise is to place them onto boards, between the folds of a linen proofing couche, with the seam underneath (à clair). The boards are then placed into fermentation cabinets called *parisiennes* for the secondary fermentation, or *apprêt*.

This is ideal, but very labor intensive because the loaves, at time of oven loading, must be transferred individually by turning them upside down onto a thin board, and then right side up onto the peel or automatic oven loader, the *tapis enfourneur*.

The loaves may also be placed onto racks with a moveable canvas carpet, or *couche mécanique*. This permits the transfer of five or more loaves in one movement, but the dough must be fairly strong because the loaves are free-standing, with no side support from the gully-like *couches*.

Another key decision in the breadmaking process is to ascertain when the loaves are ready to bake. Underproofing results in dense loaves of irregular shape—they may "explode" in places in the oven—while overproofed loaves appear bloated, have slashes that do not open properly, and have a cotton-like, disagreeable crumb.

Generally speaking, a finger imprint on a loaf that is ready to bake should disappear slowly but completely in two 2 or three 3 seconds. Imprints that don't remain for that length of time show that the dough is *underproofed*. Imprints that don't disappear mean that the dough is *overproofed*.

size cells, and the interior crumb structure tends toward greater regularity.

The cell structure of manually formed loaves is appreciably better than mechanically molded bread. The lack of proper structure is a grave fault common to mechanical molding, in spite of adaptations intended to produce improvement in this attribute. In the case of French bread, the irregular cell structure is generally very much appreciated by the consumer and may often contribute to a perceived improvement of the flavor.

THE EFFECT OF TYPE AND DEGREE OF *PÂTON* DEVELOPMENT

Under most circumstances, *pâtons* should be rather well proofed at the time they are placed in the oven.

This degree of proof should be equivalent to between 3.5 to 4 times greater than the volume of the *pâtons* just after molding.

Provided that flour quality is adequate and the proper bread production procedures have been used, the dough piece will reach final development in the oven under the effect of oven spring. The well-supported crumb structure will reach its full development under these conditions, which will tend to improve its eating qualities. Of course, this excludes the effects of any damage that the dough might have undergone from overoxidation.

With a lower proof level, the bread will be more dense, the crumb will be more or less tight-grained, and its flavor and aroma will be slightly inferior overall. Furthermore, there will be no zone of transition between the crumb and the crust, and this lack will tend to cause crust softening.

At the other extreme, when a greater proof level is achieved with the help of additives and excessive dough oxidation, the baker can load into the oven dough pieces that are six or seven times larger than their starting volume. The final development of these dough pieces in the oven under the effect of oven spring is insignificant, but it nevertheless results in loaves that are excessively large in volume. These loaves have a tendency to form "scales" or irregular flakes of crust, since the crust is very thin and fragile. Conversely, they may rapidly become cardboard-like. The crumb of such bread is very white, faded, and unattractive, with a very mediocre flavor, even to the point of actually being unpleasant.

If these overproofed *pâtons* fall upon being placed in the oven, the damage is catastrophic. The loaves are flat, the crust is light-colored, the transverse cuts do not "bloom" and spring open, and the crumb is often distasteful, with a grayish and marbled tint and a "bird's nest" cell structure. In this case, the flavor and aroma may be judged by the exterior and interior appearance of the bread: they are simply very bad.

THE EFFECTS OF FREEZING UNBAKED AND PARBAKED LOAVES

Today, production of frozen products is an important segment of total industrial bread production; this includes freezing either formed raw *pâtons*, or prebaked, fully proofed loaves. In both cases, production is carried out weeks in advance of baking, and this final step may sometimes take place at very considerable distances from the site where dough production took place.

Defrosting, proofing (second fermentation), and baking of frozen raw dough pieces are carried out at the point of sale. The necessary equipment, including storage freezer, proof cabinet, and rack oven, is known in France as a "bake-off terminal."

What effects might this production method have on the taste and flavor of bread? To respond to this question, we must distinguish between the two basic types of freezing production: the manufacture of frozen raw dough pieces, and the production of prebaked, fully proofed loaves.

In the case of loaves made from frozen raw *pâtons*, dough production procedures must take into account the following technical requirements and include the use of

- a flour with above-average strength;
- a level of yeast and additives that is two times greater than normal;
- ingredient water at a temperature of 0°C (32°F) or slightly lower;
- a quantity of water 3% to 4% less than is normal in the makeup of a dough;
- generally intensive mixing—the finished mixing temperature of the dough should be equal to or slightly less than 18°C (around 65°F);
- division of the dough pieces just following the end of mixing;
- molding of the dough pieces after just a few moments of rest (bench time);
- immediate freezing of the dough pieces in a freezing unit maintained at −40°C (−40°F). Their temperature should pass very rapidly from 20°C (68°F) to −20°C (around −4°F) at the end of the freezing process;
- storage of the frozen *pâtons* in a cold storage chamber at a temperature of minus 18°C (about −2°F), where they may remain up to 8 to 10 weeks before being used.

The frozen raw *pâtons* are delivered on demand to the bake-off location in freezer vans or trucks. They are then maintained under frozen storage at −18°C (−2°F), from which they are taken as needed. Before baking, they are placed on open mesh racks, thawed, and brought to 0°C (32°F). They are then placed in proof boxes or chambers to rise at around 22°C (71 to 72°F) and 75% relative humidity, after which they are baked in rack ovens. Since several hours are required from the beginning of thawing through proofing and baking of the dough pieces, this method does not allow the baker to respond rapidly to customer demand.

In terms of overall quality, the loaves can be regular in shape and relatively attractive if the *pâtons* have been carefully thawed and proofed and correctly scarified (slashed) before baking. Since there is no first fermentation for frozen dough and the dough is high in residual sugars, the crust is often darker than average when the loaves are taken from the oven. The structure of the crumb is generally very homogeneous. Since the degree of oxidation from the same amount of mixing is less for a cold dough (between 18 and 20°C) (65 to 68°F) than for a dough mixed at a higher temperature (normally between 24 and 26°C, (75 to 78°F), the bleaching of the dough and of the bread will be less, and the destruction of proper taste less apparent. Although there are exceptions, the use of frozen dough usually results in a crumb that has a light creamy tint. The flavor is acceptable, although the very noticeable taste of yeast is often disagreeable, since it is used in frozen dough at a rate double that of fresh dough.

In terms of overall quality and taste, bread made from frozen dough has crumb which is a desirable light cream in color, along with a highly colored crust, generally more so than common bread. An undesirable feature is that the complete absence of first fermentation deprives the dough of the organic acids that are the carriers of bread flavor, and yeast must be used at a high level. These last factors create an insurmountable obstacle to the production of good taste in bread made by this method.

The production of bread from frozen dough requires a stronger flour, greater use of yeast and additives, colder water, and an uneconomical underhydrated dough. It also requires enormous expenditures in energy for the freezing, transportation, storage, and handling of the frozen product. In the end it results in less product at a very high selling price, which is bad for everyone concerned. It is thus rather strange that ordinary bread is not always competitive with bread made from frozen raw dough.

In the second instance, the production of fully leavened, prebaked frozen loaves is usually less demanding than that of frozen raw dough in terms of storage and shelf life. Production of this product requires

- a bread flour with good strength and good baking quality;
- a slightly higher level of both yeast and baking additives in comparison to ordinary bread;
- water at normal temperature, but added at a level 2% to 3% lower than usual during mixing;
- normal mixing procedures, including the use of improved mixing, with the finished dough having a temperature around 22°C (72°F);
- normal production layout and scheduling, with provision for either an appropriate first fermentation, or the addition at mixing of 12% to 15% of prefermented dough, to save time before the division and dough forming (molding) stages; and
- a second fermentation period of normal duration after molding. The level of proof before baking should not be too great, so as to avoid loaves with excessive volume.

The distinctive aspect of this type of production lies in the baking. The loaves develop normally and reach optimal volume under the effects of oven heat. As usual, the crumb begins to take form at the same time in the loaf interior. This is the result of gluten coagulation and starch gelling, in combination with the multiplication and expansion of gas cells under the pressure of expanding carbon dioxide and water vapor. As the crust forms and begins to show the first signs of browning, the loaves are taken from the oven.

These hard rolls, half-baguettes, or baguettes are then cooled and placed in blast freezers. After freezing, they may be kept in frozen storage for 2 or more weeks, then shipped in freezer vans or trucks to baking and sales location, under the same conditions as frozen raw dough. When needed, they may be withdrawn from storage, quickly loaded onto baking racks, then immediately placed in the rack oven for finish baking. Prebaked frozen bread allows the baker to respond to the needs of the client in a very short time.

Products made by this method are nearly as expensive at retail as those made from frozen raw dough. However, partially prebaked frozen bread is of higher quality than that made from raw frozen dough, since the taste of prebaked bread benefits from a normal first fermentation or from the prefermented culture added during mixing. The major inconvenience of the prebaked method is that in the past it has been generally suitable only for small-volume bakery goods, such as hard rolls, half-baguettes, and baguettes. This was in order to counteract the tendency of partially baked products to shrink or fall. This may occur either after prebaking (from the effects of cooling and shipping), or from improper handling before finish baking. In the United States, for example, the technique known as "Brown 'n Serve" is used exclusively for dinner rolls. They are individually packaged in boxes, kept at about 0°C (32°F), and sold rapidly since they were not frozen. In general, these prebaked products are more fragile and difficult to handle and maintain throughout freezing, storage, and shipping than are frozen raw doughs.[2]

NOTES

1. This is because a longer fermentation uses more of the excess sugar.
2. "Brown 'n Serve" or "parbaked" products have recently shown a renewal in popularity in the United States, especially for specialty sourdough and hearth rolls. While these may be shipped in frozen form, they generally are shelved and sold at ambient temperature and are surface sprayed with a sorbate mold inhibitor which is similar in taste to the natural sorbates in some sourdoughs. Very recent developments in vacuum cooling of specially processed parbaked products in the United Kingdom (the Milton-Keynes process) are said to have extended the use of parbaking procedures to all types of bakery products, and appear to have eliminated the problem of product shrinkage while reducing or negating the need for frozen transport and storage. Whether this process will be successful in other English-speaking countries remains to be seen.

PART III

Baking and Keeping Qualities of Bread and Their Relationship to Taste

Bread Crust

Chapter 7

- Ovens used in bread baking.
- Formation, coloration, and degree of crust baking and their relationship to bread taste.
- The effects of oven steam on crust taste.
- Flour-dusted breads and crust taste.
- Scaling of bread crust.
- Frozen storage of baked bread.

OVENS USED IN BREAD BAKING

The baking of dough and its transformation into bread is the most important stage in breadmaking, and it has a great effect on the taste of the final product. Baking may take place in any of several types of bread ovens, including

- masonry ovens, which are simply an improvement of the stone oven known since very ancient times. These are very rare today. They may be wood burning (Figure 7–1), with combustion occurring either directly on the **sole** or floor of the baking chamber or in a firebox located in the forepart of the oven. In the latter case, the heat, flames, and smoke from the burning wood are pulled into the heart of the oven by the effect of chimney draw or draft. Gas burners may be used in the same manner. Although fuel oil was also in common use for several decades, this is no longer permitted in France. Multiple baking cycles may be carried out after oven heating as a result of the accumulation of heat energy in the masonry mass of the oven, which generally weighs between 40 and 50 tons.

Heating is by means of burning wood on the oven floor; the glowing embers and cinders are raked from the oven mouth. The oven chamber is cleaned with an old, water-soaked jute sack fastened to a long wooden handle—the oven swab—to remove the remaining cinders. This operation is assisted by a strong air current, which draws dust and odors toward and through the oven flue, releasing them into the atmosphere.

When these ovens are properly heated, the decreasing temperature curve is well suited for baking bread, especially large loaves. Such loaves are placed in a well-heated oven, around 250°C (480°F), and the interior temperature of the oven decreases gradually, while the baking process advances inversely at a comparable rate. The crust forms gradually while the outer layer of dough releases its moisture, and this progressive change allows proper baking and avoids the risk of a burned or overly browned crust.

Some bakers have advanced the opinion that bread baked in ovens heated by burning wood directly in the baking chamber possesses an aroma and taste that are derived from the burned wood. Provided that the baking chamber has been cleaned properly following oven heating, this is simply a marketing ploy—a myth that has commercial success as its only real basis. However, this marketing myth should in no way detract from the excellent baking qualities provided by this type of oven.

- continuous baking ovens with a rotating circular sole and indirect heating. It is possible to heat these ovens with gas jets or fuel oil burners, but in actuality they are nearly all wood-fired. These ovens possess a very high level of thermal inertia and, once they have been brought to operating temperature, require relatively little fuel while providing excellent baking conditions. For a time they were credited with all the aroma and taste benefits associated with wood heating. Since the oven chamber is airtight and separated from the heat source, it is obvious that these virtues are nonexistent and really only a waste of words.
- continuous bake ovens with fixed oven deck and indirect heating (Figure 7–2). These ovens may be heated with either gas or fuel oil by recycling the heated air and combustion gases, or by vapor tubes that surround the baking chamber. In both

Baking loaves directly on the hearth is ideal for producing proper crust and interior structure.

Figure 7–1 Wood Burning Oven. Copyright © Jules Rabin.

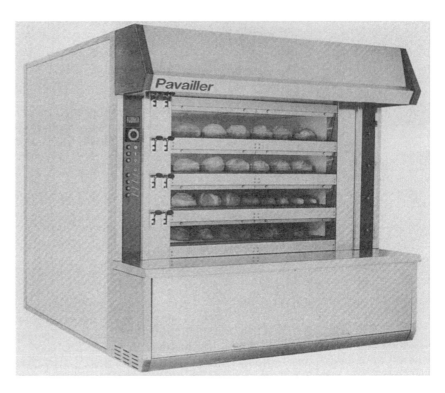

Figure 7–2 Oven with Fixed Oven Deck. Courtesy of Pavailler, Inc., Dorval, Quebec.

Figure 7–3 Rack Oven. Courtesy of F.B.M. Baking Machines, Inc., Cranbury, New Jersey.

The appeal of wood-burning ovens is undeniable, but if properly cleaned and operated they have no actual influence on bread flavor. In France, it is illegal to claim that a wood-burning oven was used in bread production when an indirectly heated oven has been used.

In reference to the use of wood-burning ovens in general, Professor Calvel has written that "in order to perceive a difference in flavor between bread baked in a wood-burning oven and that baked in an appropriate deck oven, one must be blessed with taste buds equipped with a fertile imagination."

Rack ovens with screens are a very practical method of dealing with many loaves with a minimum of handling. However, the gain in efficiency comes at a price, because the bottom crust and the crumb structure are both adversely affected.

cases, the heat source is completely independent of the baking chamber.

These ovens are composed of several stacked baking chambers. Today, the oven is loaded with dough pieces with the help of machinery, i.e., retractable canvas belt oven loaders, which have generally replaced manual oven loading with a baker's peel.

All of these ovens provide good baking conditions. Because of their low thermal inertia they are better suited to the production of long loaves (baguettes and 400 g loaves) than of thick or large loaves. However, these latter types may be successfully baked in deck ovens, provided that proper precautions are taken.

Electric ovens are encountered more and more frequently in this general class. They are compact, offer unparalleled ease of use, and have notable hygienic and environmental advantages.

- rack ovens (Figure 7–3). These ovens are in the form of a cabinet equipped with a rotating baseplate. They are heated by recirculating hot air within the internal walls of the baking enclosure. The fuel for heating may be gas, fuel oil, or electricity.

In use, the baseplate holds a movable rack equipped with 12 to 18 trays or frames. These frames may be equipped with metal screens or a series of long, round-bottomed baguette forms onto which the dough pieces are deposited.

These screens or forms are generally made of stainless steel and coated with nonstick silicon resins. This nonstick coating prevents the dough from sticking to the screen, but the screen must be periodically removed and recoated.

The baking conditions in rack ovens are different from those in soleplate ovens. In the case of soleplate (deck) ovens, the heat exchange between the oven and the dough takes place from the very beginning of the baking cycle according to the three principles of heat transfer: conduction, radiation, and convection. With rack ovens, heat transfer takes place at first only by convection. Radiation and especially conduction occur with a slight delay. This brings about a degree of rigidity in the exterior layer of the *pâton*. The *pâton* also becomes attached to the metal net of the baking form. As a result the structure of the loaf is less well developed than it might otherwise be, since oven spring is less intense than usual.[1]

The result is baking conditions that are less favorable to large round or tall loaves than to lighter, long ones. The crumb of bread baked in a rack oven often has a finer cell structure than bread baked in other types of ovens. The bottom crust is thinner and more rubbery, with a greater tendency to become soft. The consumer often has the impression of biting into a piece of surgical gauze.

The flavor is also perceived differently, as it is somewhat inferior to that of bread baked in fixed sole ovens. However, it would seem that this quality factor is not always taken into consideration by the consumer.

- high-volume industrial baking ovens. These are tunnel ovens for the most part, some of which have several stages or zones, heated by steam transport tubes that surround the baking chamber. These are indirectly fired continuous baking ovens and may be heated with gas, fuel oil, or electricity. Heat transmission is either by means of recirculated air or by steam tubes, and the baking conditions achieved are excellent.

FORMATION, COLORATION, AND DEGREE OF CRUST BAKING, AND THEIR RELATIONSHIP TO BREAD TASTE

Bread baking results from the transfer of heat between oven and dough. In the course of this process, the *pâton* undergoes numerous changes that may be grouped into three important stages:

1. During the first stage, there is a strong evaporation of moisture from the exterior layers of the *pâton*. This is mitigated by injection of steam into the baking chamber just before oven loading and by the resulting condensation of moisture onto the cool surface of the dough piece. As a result of this cycle of condensation and evaporation, the temperature rise at the surface of the *pâton* is slowed, and the surface retains a degree of elasticity that aids further loaf development and expansion. The evaporation results in a massive consumption of heat, despite the elevated temperature of the oven (from 240 to 250°C or 465 to 480°F) and the good conductive qualities of dough because of its high moisture content.

At the same time, there is a rise in the internal temperature of the *pâton*, resulting in a period of intense fermentation. This causes an acceleration in the production of carbon dioxide gas, and individual gas cells in the dough undergo a rapid expansion. Bakers refer to this as the initial "oven spring" effect, which produces spectacular size development of the dough piece. Fermentation continues to take place until the internal temperature of the *pâton* reaches 50 to 60°C or 122 to 140°F. Yeasts are killed by this temperature

Figure 7–4 Slashing Loaves (Scarification).

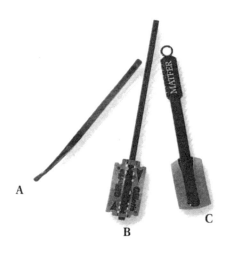

Figure 7–5 Cutting Blades.

The proper slashing of loaves (Figure 7–4) is an important step, and must be done with a suitable blade just prior to baking (Figure 7–5). Traditional blades (A) are available in various shapes. Professor Calvel always travels with a selection of these, as well as a miniature sharpener. Many bakers slip a double-edged razor blade (B) over a traditional blade or over special metal holders that curve the blade slightly, but this practice has been forbidden by law in France following a small number of excruciatingly painful consumer mouth injuries. Blades in a special plastic sheath (C) are now in use. Proper results cannot be obtained by use of inappropriate tools, such as serrated knives or scissors.

Slashing or Scarification of Loaves

For best results, loaves must be slashed at only a slight angle to approximately 1/8″ depth (Figure 7–6), with each slash beginning to the right of the last one third of its predecessor. The overlapping slashes expand and open out into lozenge shapes during the early stages of baking. (Figure 7–7 shows proper and badly slashed baguettes.)

Slashing the loaves at an angle creates a raised lip as the loaves bake, an especially crispy strip of crust which is called the *grigne* in French.

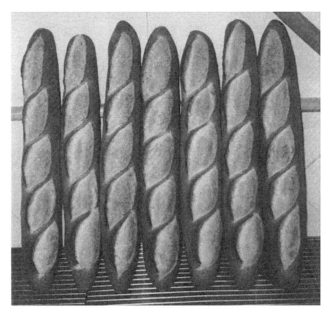

Figure 7–6 Beautiful loaves, perfectly slashed. Courtesy of Lesaffre Ingredients, Marcq-en-Baroeul, France.

Figure 7–7 Properly versus Badly Slashed Baguettes.

rise, which marks the end of carbon dioxide gas production, and the end of the first stage of baking.

Especially in the case of French bread, this volume development is enhanced by the cuts made on the dough surface before oven loading. These incisions facilitate oven spring and increase the development and volume of the *pâtons*. In addition, they also improve the exterior appearance of the baked bread because of the regular swellings that result, which encourage the growth of gas cells within the crumb.

2. During the early second stage, the still-plastic dough continues to expand under the combined effect of the increasing expansion of carbon dioxide gas and the beginning of steam formation within the *pâtons*. The simultaneous temperature increase progresses steadily toward the center of the dough piece and brings on both starch gelatinization and gluten coagulation. As the internal temperature reaches about 70°C (158°F), the plasticity of the *pâton* and its development are brought rapidly to an end. This marks the end of the second stage, and the developing loaf has attained its final volume.

3. In the third baking stage, the high level of evaporation from the external loaf surface diminishes, while the surface temperature increases and the crust forms and thickens. The residual sugars that remain in the dough are caramelized, and nitrogen-bearing substances undergo a Maillard reaction. Both of these surface reactions contribute to the formation of crust color. The rate of evaporation from the exterior loaf walls decreases as this stage progresses, and the need for intense oven heat decreases as well. For this reason, it is advisable that the oven temperature curve decline toward the end of this third stage. Finally, it should be noted that the interior temperature of bread hardly reaches 100°C (212°F) during baking,

Figure 7-8 Making *Fendu* Loaves.

An alternative to slashing long or round loaves is to "split" them with a rolling pin (Figure 7-8). The resulting "trough" opens somewhat during baking. These "fendu" loaves are not split into two pieces. Shortly after shaping, the loaves are sprinkled with flour, and a rolling pin is used to form a trough by using a firm back and forth motion. The two halves of the loaf are connected by a "hinge" of dough approximately $1/10$ or $1/12$ the thickness of the unbaked loaf. After the central "split" is pressed into the dough, the dough pieces are placed upside down into linen-lined baskets called *bannetons* to rise, and then turned right side up just before baking.

and almost never goes beyond that point, while the exterior temperature of the crust rises to an average of 225°C (437°F).

In discussing bread crust and its taste, INRA engineers D. Richard-Molar and R. Draperon indicate that "at the surface of the dough piece, the temperature may reach from 230 to 250°C, and from 180°C upward, nonenzymatic Maillard and caramelization browning reactions gradually form the crust. All of these reactions are accompanied by the production of characteristic volatile composites.

When heated above their point of fusion, the sugars that remain on the surface of the dough piece produce colored degradation products that have an acid taste, or even a slightly bitter or astringent (such as hydroxymethyl or furfuraldehyde) taste, and a series of composite carbonyls, aldehydes, and keytones with agreeable volatile odors. These caramelization reactions are still poorly understood, and we have little data concerning the ways in which they occur.

Maillard reactions have been the subject of numerous studies, and are better understood, in spite of their great complexity. In the first stage, an additive compound is formed between a reducing sugar and an amino sugar. Through a series of reactions, this composite gives rise to amino-dioxyketose, which undergoes decomposition either by dehydration or by splitting.

Dehydration produces very highly scented furfural compounds through the further loss of three water molecules. If only two water molecules are lost, very highly reactive composites called dehydroreductones are formed. By reaction with 2-amine molecules, they form aldehydes based on the amine molecule. This so-called Strecker reaction may also involve carbonyl compounds, which are formed by fermentation or by the oxidation of fats. The Maillard reaction may be the origin of heterocyclic nitrogen compounds—pyrazines for example—which exhibit very intense odors, such as that of toast in particular.

In addition to these basic changes, less important factors in baking have notable, but less crucial, influences on the taste and flavor of bread. These include

- the protein level and enzymatic content of the flour,
- the composition of the dough,
- the condition and developmental level of the *pâtons* at the time of oven loading,
- the type of oven and temperature level,
- the presence or absence of steam in the baking chamber, and
- the duration of baking.

All of these factors have an effect on the nature of the crust, its thickness and degree of coloration, its dryness and crispness, and its crumbliness and fragility.

A flour that is low in protein or deficient in enzymatic activity or an oven temperature that is too low will generally result in a pale, often relatively thick and hard crust. There will be an inadequate Maillard reaction and insufficient caramelization, and as a result the taste of the crust and of the bread will be inferior—insipidly tasteless and in general not very agreeable. On the other hand, a dough with a protein level or sugar content that is too high or that is baked at too high a temperature generally produces a crust that is too dark and thin, with a tendency to become soft. The taste of the bread will be more pronounced, but the flavor will be less than pleasant. The pleasant texture perceived by the mouth will be reduced, the crust will be less crisp, and in general the flavor will be less pleasant.

The degree of crust coloration and its thickness have an important bearing on both the taste of the

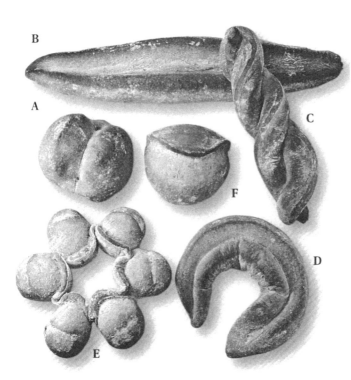

Figure 7-9 Various Country-Style Loaves.

Figure 7-9 shows various country style loaves, including: (A) boule fendue, (B) marchand de vin, (C) tordu, (D) couronne fendue, (E) couronne rosace, and (F) tabatière.

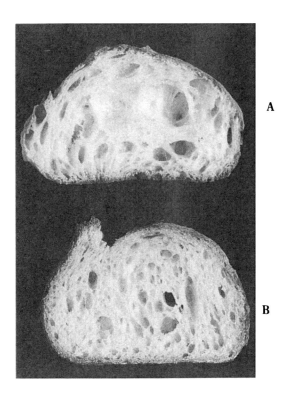

Figure 7-10 Cross Sections. (A) Rustic Bread, (B) Sourdough (pain au levain).

∎

Firm, but gentle forming results in a crumb structure that is characterized by large, irregularly shaped and sized holes or cells. This crumb structure is a distinctive characteristic of authentic French breads (Figure 7-10).

crust and that of the whole loaf. It is the task of the baker to respect the delicate equilibrium between the sugar and protein contents of the flour. He or she must also regulate both the temperature of the oven and the amount of baking time in order to achieve the optimal crust coloration. Such a crust will possess the optimal combination of true bread aroma—a subtle amalgam of toast, hazelnut, and frying odors—while avoiding the excessive coloration that results in a bitter taste. Proper bread color must not be excessive, but should be just slightly darker than golden yellow, with a discreet orange tint that borders on light brown.

To avoid an insipid crust taste, it is extremely important that the crust color not be lighter than golden yellow. If this occurs, there has been an inadequate Maillard reaction, associated with incomplete caramelization, both of which will noticeably affect the aroma and taste of the bread. In those areas where the bakers have customarily responded to consumer demands for lighter crust coloration—for example, in southeastern France, Italy, and Brazil—it is desirable to try to inform the consumer about this problem, to give him or her the opportunity to make a comparison, and to encourage the consumption of a more highly colored and flavorful product.

The type of oven may also have a certain degree of influence on bread crust flavor, although its effects are less apparent than the other factors discussed above. Intermittently heated masonry ovens, with their high degree of thermal inertia and a gradually declining baking temperature curve, tend to produce bread with a thick crust and good keeping qualities. The crust is relatively easy to chew and not very crisp. Continuously heated soleplate ovens produce breads with slightly thinner but more crispy crust, although the keeping qualities are more limited. Pulsed-air rack ovens produce a thin and rubbery bottom crust with a tendency to quickly lose any crisp qualities and become cardboard-like—a notable deficiency. In summary, these differences may be lessened by proper adjustment of the oven, although this effect is not always apparent.

THE EFFECTS OF OVEN STEAM ON CRUST TASTE

In comparison to the influence of oven type, the presence or absence of steam in the baking chamber during the first part of the baking cycle has a much

Figure 7-11 The Unpleasant Flavor of a Burned Bottom Crust Migrates throughout the Loaf.

∎

Proper oven temperature is determined by the size and density of the loaves to be baked, and proper baking involves simultaneously achieving desired crust color and crispness. High oven temperatures lead to browning before the crust is formed, which leads to crust softening after cooling.ABnormally fast browning, or slow browning, forces the baker to reduce oven temperature and bake longer, yielding a thick, dry crust. This may be caused by hyperdiastatic or hypodiastatic flours.

Special attention should be paid to the bottom crust, because the unpleasant flavor of a burned bottom crust migrates throughout the rest of the loaf (Figure 7-11). It is important to avoid moving the loaves as they bake in the oven, because this is one of the major causes of crust burning.

Figure 7-12 Various Sizes, Shapes, and Slash Patterns of French Breads.

Figure 7-12 shows sizes, shapes, and slash patterns for a variety of French breads. The varying weights and lengths of long loaves give consumers a choice of textures and ratios of crust to crumb. Table 7-1 provides information regarding characteristics of typical French bread types.

The *épi de blé* or wheat stalk loaf is considered by many to be an attractive loaf, but the cutting of the unbaked loaf with scissors penalizes the texture.

Ficelles, which are mostly crust, do not show French bread to its best advantage, because it is the flavor and texture of the interior of the loaf that sets great bread apart from ordinary or mediocre bread.

By the 1920's, factors permitting the production of lighter, crispier loaves were in place: stronger flours, kneading machines, yeast-only formulae (i.e., no sourdough), steam-injected ovens. Parisian customers were fascinated by the new style, and the bakers' answer to them was to produce longer, skinnier loaves with a higher crust-to-crumb ratio. The baguette and other loaves were collectively called 'Pains de Fantaisie' not only because they were lighter, more delicate and more expensive, but also because they had to be eaten on the same day.

Table 7-1 Typical French Bread Types and Their Characteristics

Loaf Type	Unbaked Weight	Baked Weight	Number of Blade Cuts	Length
Baguette	350 g (12.35 oz)	250 g (8.82 oz)	7 cuts	60 to 70 cm (23.6 to 27.56 in)
Ficelle	185 g (6.53 oz)	125 g (4.41 oz)	5 cuts	40 to 45 cm (15.75 to 17.72 in)
Long bâtard	350 g (12.35 oz)	250 g (8.82 oz)	4 cuts	35 to 40 cm (13.78 to 15.75 in)
Short bâtard	350 g (12.35 oz)	250 g (8.82 oz)	3 cuts	30 cm (11.81 in)
Parisian	550 g (19.4 oz)	400 g (14.11 oz)	5 to 7 cuts	60 to 70 cm (23.6 to 27.56 in)
Petit pain	75 g (2.65 oz)	50 g (1.76 oz)	1 to 3 cuts	8 to 10 cm (3.15 to 3.94 in)
Boulot	350 g (12.35 oz)	250 g (8.82 oz)	1 or 2 cuts	20 to 25 cm (7.87 to 9.84 in)

Source: Data from Hubert Chiron, INRA, Nantes, Fance.

Figure 7–13 Too Much Steam, No Steam, and Correct Steam.

greater impact on the taste of bread (Figure 7–13). Bread may be baked without steam, just as it was almost exclusively until the beginning of the 20th century. In that era bakers primarily made large round or dense, compact loaves, and the oven development of those bread types suffered less from its absence than would the longer and lighter loaves made today.

Without the advantage of steam, bread crust takes on an earthy brown tint, and its surface is dull and uneven. This is what is commonly known as "gray bread". While the flavor is somewhat pleasant, with a dominating note of toast mixed with a very slight bitterness, the overall taste seems to be a little "flat".

With the steam injection method developed during the 1920s, however, the formation and coloration of the crust are very much different. As a result of the supersaturation of the baking chamber with steam, moisture condenses on the cooler surface of the dough piece during the first moments of baking. This phenomenon facilitates the expansion and development of the dough piece and also has the effect of slightly diluting the surface starch. Subsequent gelatinization creates an "icing" effect, which smooths the crust and gives it a very attractive appearance after browning. The result is a fine, thin, more crisp crust, with a more pronounced hazelnut flavor that is very appetizing and pleasing. Nevertheless, this finer crust has an unfortunate tendency to soften more rapidly in humid weather.

FLOUR-DUSTED BREADS AND CRUST TASTE

The baking of flour-dusted breads is a fad that has grown significantly since the 1980s. Up until that period, flour dusting of breads was limited to dense, compact loaves, especially those called "country-style" breads. Today this practice extends to a significant part of the overall production of longer breads, and even to baguettes.

The practice has been handed down to us from ancient times, when the majority of bread produced was large loaves. The dough pieces from which these breads were made were not coated with flour for the purpose of producing flour-coated loaves—with the exception of one type of long loaf called "polka" bread. They were instead lightly dusted with flour to prevent the dough from sticking to the sides of the wicker proofing baskets—which may or may not have been canvas lined—during the long fermentation periods necessary to the production of these naturally cultured breads (i.e., sourdoughs).

Since these dough pieces were generally baked without steam, they were "gray" loaves. The remaining superficial traces of flour remained visible and gave a slight (or very slight) flour dusted appearance to the crust. Today, the reason for flour dusting is quite different. The practice is intended to give the baked bread a more rustic or "traditional" appearance. By evoking the countryside, it suggests to the consumer a "back to nature" ideal. If I were of a critical turn of mind, I would call this the exploitation of the make believe.

Among these flour-dusted breads one finds country-style loaves, naturally leavened breads, long loaves, the previously noted polka breads, and even *bâtards* and so-called country-style or specially produced baguettes. These types of breads are generally sold at higher prices than authentic traditional breads. It sometimes happens, however, that the quality is not higher. The crumb of the bread is white, without aroma, and devoid of taste as a result of overmixing and oxidation. I distinctly remember writing an article on this subject as far back as the 1960s, with the rather literary title "Country-style breads, or the flour-dusted travesty?"

Having said all that, I must ask: "What is the real reason for flour dusting?" If we disregard the marketing value of flour dusting, which is not a small consideration, what valid reason remains? At the risk of disappointing both professionals and consumers, I must say that, in my opinion, there is very little reason. If the flour is applied in a very thin layer and is singed to a light reddish brown color—which does not always happen—it will impart only a very light "singed" taste. It must be admitted that this will be very slight and therefore without any great importance. If the layer is applied more thickly, the flour will remain nearly white after baking and will in no way improve the flavor of the crust. I would even maintain that the opposite is true.

It is important for the baker to take special care to avoid applying too heavy a layer of flour to the *pâton* (Figure 7–14). When applied in a thick layer, the flour acts as a screen, protecting the exterior layer of the dough piece from heat radiation and inhibiting both baking and crust coloration. It does this so effectively that when the layer of flour is removed, the crust of the bread is found to be practically white and the taste is greatly diminished, since caramelization has not taken place. When caramelization is properly controlled, this complex process of coloration produces the bouquet of tempting aromas in the crust and enriches the flavor of the crumb and of the entire bread by their diffusion.

It should also be noted that dough pieces cannot be dusted with flour without creating some hygiene problems. When the dough pieces are placed in wicker baskets or on flour-dusted canvas, the flour absorbs moisture from its contact with the dough. The canvas rapidly becomes soiled, promotes the growth of mold, and offers an ideal home for mites. Meticulous cleaning is thus required. When the flour is applied with a sifter before loading the oven, both the oven and surrounding area require careful and conscientious upkeep.[2]

When an honest evaluation is made of the advantages and disadvantages of the production of flour-dusted breads, the disadvantages clearly outweigh the advantages. While flour dusting may be attractive, it is also bothersome, and the flour can sometimes stick to the roof of the mouth! If the baker decides that he or she absolutely must dust loaves, it is in his or her best interest not to overdo it, so that the resulting travesty will not be harmful to crust coloration and to the development of the rich aromas it carries.

SCALING OF BREAD CRUST

The appearance of crust scaling as a bread defect coincides with overmixing, hyperoxidation of doughs, the overuse of certain additives in the flour or the dough, and the exaggerated volume of the bread that is the result of such practices. It should be remembered that when bread is removed from the oven after baking, it passes quickly from the temperature of the oven around 225°C or 437°F to the ambient temperature of the bakehouse or shop, which is generally between 25 and 30°C (77 and 86°F). The crust thus undergoes a violent thermal shock on the scale of 200°C.

In the few moments that follow oven discharging, the loaf contracts and the crust begins to develop small cracks and crevices. When the loaf is of medium size or rather small, as was the case before the early

Figure 7–14 Properly and Overly Floured Loaves.

1960s, the crust is generally thicker and better able to resist this shock. The cracks are generally small and without any real importance. On the other hand, when the loaf is very high in volume, the crust is thinner and more fragile. Deeper cracks appear in a much more dense pattern, which encourages the development of small scales. As the bread cools, these scales are less and less firmly attached to the loaf.

It is for this reason that whenever large volume loaves are transported or shaken, small scales of crust become detached. At its most serious extreme, when such loaves are sliced, they undress shamelessly! This is an unfortunate effect that is very little understood by the consumer, but it occurs very often in France, Brazil, and certain other countries of Latin America.

As a result, the consumer accuses the baker of having sold bread that has been frozen after baking. Such a thing does happen, and it is logical to think that scaling would be aggravated by the bread having undergone two additional thermal shocks. That is rare, however, and in any event scaling would already have been preordained by other causes.

Excessive scaling has its root cause in the excessive volume of bread and the nature of the thin and fragile crust that surrounds it. In turn, the exaggerated volume of bread results from excessive development of the gluten network and from hyperoxidation of the dough, both of which are the logical consequences of overmixing. The presence of ascorbic acid is small comfort for France, since when it is overused the amount of mechanical work applied to the dough tends to create a drier crust that is even more prone to scaling. A dough overly rich in protein also tends to encourage scaling.

Here again it may be said that excess in anything is a defect. The practices of enriching the flour with lipoxygenase, dough overmixing and overoxidation, and the excessive use of ascorbic acid all combine to yield poor-quality breads. They are too large in volume, with a defective crust that is often fragile, cracked, and cardboard-like, and a crumb that is snow white, inconsistent, cottony, and without taste. All of these represent a set of circumstances fatal to the quality and consumption of French bread, and indeed for all breads in general.

FROZEN STORAGE OF BAKED BREAD

As noted previously, the practice of freezing may serve to aggravate the tendency of bread crust to scale. However, bread may be protected from scaling by taking a few simple precautions, and freezing may be used to advantage, or at least without damage to the product. It is necessary to observe the following rules:

1. Freeze only fresh bread.
2. Protect the loaves from the atmosphere of the freezer or cold storage enclosure by placing them in air- and moisture-proof wrappers either before, or preferably immediately after, they have been frozen.
3. Limit storage as follows:
   —3 days maximum for baguettes and long breads
   —7 to 8 days for pan breads and compact, dense-crumb loaves
4. Thaw them carefully, either by placing them in a warm and humid environment and then allowing them to defrost further under ambient conditions or by allowing the thawing to proceed entirely under ambient (room temperature) conditions.[3]
5. In all instances, protect the loaves from air movement, which encourages drying of the crust and consequent scaling.

NOTES

1. Professor Calvel's French printer may have left a few words out of the original sentence. However, reduced oven spring is common with rough-surfaced baking forms to which the dough adheres.
2. This is important for hygiene and worker safety. Flour dust on a tile or sealed concrete floor can be extremely slick and may cause serious falls and consequent injury.
3. Some researchers have found that accelerated staling occurs at the critical zone just above freezing, so it is advisable to bring the loaves through this range as rapidly as possible to minimize staling. This also explains why bread should never be refrigerated.

Bread Crumb

Chapter 8

- Formation and baking of the crumb.
- Crumb color and cell structure.

FORMATION AND BAKING OF THE CRUMB

Bread crumb is formed by the action of heat on bread dough, just as in the case of the crust. After the dough has been brought to full development by multiplication and expansion of the entrapped carbon dioxide gas cells, the heat of the oven raises the dough temperature to a level that causes gelatinization of the starch and coagulation of the gluten. This gives the crumb its final structure. Accompanying this change, but at a slightly higher temperature, the amylases present in the dough are destroyed.

The baking of the crumb also has an important influence on bread taste. Researchers D. Richard-Molar and R. Draperon expressed the opinion that since the internal temperature of the dough continues to rise until it stabilizes at only 100°C (212°F), the contribution of the formation of new substances as a result of heat effects is relatively limited. They continued their discussion to state: "However, it is necessary to bear in mind the splitting of the hydroperoxide of lineoletic acid, which is formed by the action of lipoxygenase. This occurs when intensive mixing is used in conjunction with bread flour which contains added bean flour, and results in an intensified production of hexanol in the crumb." (Hexanol is non-typical odor in bread and is reminiscent of rancidity.)

The work described by R. Draperon in 1974 led them to add that

> in the case of a breadmaking method which included a conventional mix stage, the formation of these volatile components is very limited, since the lineoletic acid is only partially oxidized. On the other hand, when intensive mixing is used in conjunction with the addition of bean flour, this acid is completely transformed into hydroperoxide. Under such conditions, the amount of hexanol produced in the bread crumb during baking is **ten times greater** than the amount produced in the crumb of conventionally mixed bread. The production of other volatile substances resulting from different types of oxidation reactions might also be increased under intensive mixing. These volatile substances, and especially hexanol, may spoil the taste of the bread by changing the equilibrium of the typical aroma components.

Thus, we can clearly see that bread taste may be influenced by overmixing and the resulting hyperoxidation of the dough. It may also be said that even if the crumb is only slightly enhanced by the creation of new volatile components under the effect of the heat of baking, it nevertheless remains clear that the crumb is the reflection of

- the nature and type of flour,
- the type of leavening system,
- the degree of mixing intensity,
- the type of loaf forming technique(s), and
- the kind of oven used.

I would even go so far as to say that the aroma components of the crumb are more important than those of the crust because the crust, no matter what type of dough is involved, always benefits from the effects of the Maillard reaction and the caramelization of the residual sugars present in the dough. These two effects tend to "even out" differences in crust quality. It is especially the crumb that makes the difference in terms of aroma, taste, and flavor between a good and bad bread.

How is it possible to arrive in advance at an opinion on bread quality through a simple touch and visual test, and to then make a favorable or unfavorable judgment? It seems to me that for French bread, and indeed for bread in general, three criteria may be useful to the consumer as a guide in identifying the crumb of a high-quality product: the color of the crumb, its structure, and its plastic properties.

CRUMB COLOR AND CELL STRUCTURE

Crumb Color

Color is the criterion that is most evident. In France, a type 55 is the normal flour, and it is milled to have an ash level between 0.50 and 0.60, but preferably closer to 0.50. A good bread—a real quality loaf—made from this flour will have a creamy white crumb. This tint will be homogeneous across the entire surface of a slice of this bread, whether it was yeast-leavened by the straight dough method or made from a prefermented culture.

An irregular tint that includes darker shadow zones is called "marbled" and is the obvious symptom of dough that suffers from a lack of maturation or a lack of strength during forming. A more resistant, less elastic crumb is characteristic of these shadow zones.

The proper creamy white color of the crumb shows that dough oxidation during mixing has not been excessive. It also presages the distinctive aroma and taste that are a subtle blend of the scent of wheat flour—that of wheat germ oil, along with the delicate hint of hazelnut aroma that comes from the germ. All of these are combined with the heady smell that comes from alcoholic dough fermentation, along with the discreet aromas that are the results of caramelization and crust baking.

When the dough has been leavened with a natural *levain* or a *levain de pâte*, the crumb will be somewhat different. It will still be a light cream color, but the creamy tint will be slightly less evident and will include a very light grayish cast. Here again, the color should be homogeneous across the entire surface of a slice of this bread. If marbling is evident, one may assume that the same cause and effect apply here as well, and that there was a lack of dough strength at the forming stage. Since this type of bread is usually more dense, the shadow zones of the crumb are darker, more resistant, and a little less elastic than with yeast-leavened breads. The baker must be sure with this method that the maturation and strength of the dough have reached optimal levels at the time of loaf forming.

Cell Structure

This subject has already been touched upon in discussing the production and forming of the *pâtons*. It was noted that the original and proper texture of French bread is characterized by a notably irregular gas cell structure. In traditional French bread, a regular and homogeneous structure is indicative of a gummy crumb, with inconsistent eating qualities that tend to detract from any flavor the bread may possess. This is especially true in the case of large, high-volume breads.[1]

On the contrary, the grain of French bread should be open, marked here and there by large gas cells. These should be thin-walled cells, with a lightly pearlescent appearance. This unique structure, resulting from the combination of numerous factors including the level of dough maturation and the loaf forming method, is basic to the eating qualities, flavor, and gustatory appeal of French bread. In summary, it may be said that a well-structured, supple, elastic crumb with a passably open cell structure and a light cream color with pearlescent highlights is the indispensable mark of a truly high-quality French bread.

NOTE

1. It should be noted that this is not true in the northern part of France, where the custom of making snacks or open-faced sandwiches with sliced bread made the production of an evenly textured bread necessary (so the fillings would not fall through the large holes typical of most French breads). This custom is not quite as popular as in the past.

Bread Staling

Chapter 9

- Storage and staling effects on bread taste.
- Bread staling and factors that influence it.
- Consumption of stale bread.
- Shelf life and taste of industrially produced packaged breads.
- Types of bread spoilage.

STORAGE AND STALING EFFECTS ON BREAD TASTE

Bread enters a state of constant change after baking. As soon as it leaves the oven, it undergoes rapid cooling. Throughout this stage, alcohol vapors, carbon dioxide, and especially water vapor are diffused through the crust and replaced by ambient air. Water vapor may condense on cold surfaces when it is trapped between them and the crust with which they are in contact, and the bread is said to "sweat." This is the origin of the term *ressuage* to denote the cooling period that occurs after baking. The duration of this cooling period is a function of the size of the loaf; it may last from 30 to 90 minutes, and sometimes even longer.

Because of the high water content of the crumb during this period, it may easily be formed into balls or pellets with finger pressure. Some consumers view this as proof that the bread has been improperly baked. This statement is generally false. It should be taken into consideration that the crumb retains almost the same moisture as the dough before baking—within about 1%. After baking, practically all types of bread crumb may be formed into balls because of the presence of so much moisture. Figure 9–1 represents the distribution of moisture within the crust and crumb of French bread.

It would seem that bread is most appetizing when it is still warm or only partially through its "sweating" period. About two-thirds through the cooling stage, the crust is pleasantly scented, crisp, and very appetizing. At this same stage, however, the odor of the crumb is less engaging. Furthermore, when chewed it is markedly doughy, with a tendency to form a mass in the mouth. The digestive juices are able to penetrate it only partially, and bread consumed at this stage may be digested only with some difficulty.

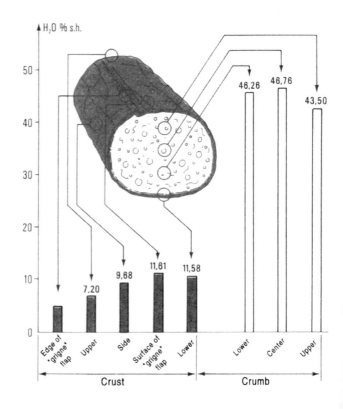

Figure 9–1 Distribution of Moisture across a Cross-Section of French Bread.

The point at which bread is most acceptable is when equilibrium has been reached between the diffusion of aromas from the crust toward the crumb, and vice versa. This optimal point is reached when the bread is practically cold, and the ressuage, or cooling cycle, is nearing its end.

The crust has a brilliant shine, still possesses a fine aroma, and has a very agreeable texture when eaten.

The creamy white cell walls of the crumb are often semi-transparent, with a pearly iridescence, and give off a rich aroma in which the odors of wheat flour are blended with those produced by alcoholic fermentation. Such bread is fresh and mellow to the taste and pleasant to chew. Whenever a loaf has reached this stage of cooling, and has been made in a manner that respects the fundamental nature of things, one truly experiences the very essence of fresh bread while eating it.

Bread should receive special care during the cooling period: during cold or dry weather, it should be protected from air currents to avoid excessive drying. In humid weather, it should be kept in a well-ventilated area to keep the crust from becoming overly soft. It should be kept under temperate conditions, about 24°C (75°F) to avoid shock and premature staling from excessive cooling. If these reasonable precautions are taken, long breads, baguettes, and 400 g loaves may be kept fresh and in excellent eating condition for between 15 and 24 hours, and compact round loaves and heavy large ball or round loaves may have a shelf life of 2 to 3 days.

BREAD STALING AND FACTORS THAT INFLUENCE IT

Bread continues to change even after it has gone through a cooling period. It continues to become more dry and to lose moisture, except during very humid weather. The drying of bread should not, however, be confused with staling, although it does facilitate it.

The staling phenomenon, which takes place from the time of baking and throughout the storage period, results for the most part from changes in starch structure. Starch undergoes gelatinization, which leads to the "starching" effect, under the effect of oven heat. In this state, starch links with and is deposited onto coagulated gluten to form the crumb structure. At this stage it is in an amorphous and unstable state.

The crumb that results from the combination of starch and gluten remains relatively soft, elastic to the touch, and mellow in the mouth after cooling and for several additional hours, while the crust remains crunchy. As time passes, however, the starch gradually changes to a crystalline state, reverting to its original form. This change is evident in the progressive hardening of the crumb and the corresponding loss of softness and elasticity, as well as a lessening of its flavor.

It should be pointed out that the starch element of staling is partially thermo-reversible. Stale bread recovers many of the properties of fresh bread when reheated, although for only a short time. Note also that after coagulation the gluten begins to form a gel, which also begins to evolve irreversibly to a harder state. This occurs less rapidly than in the case of starch, which is why a tendency toward slowing of the staling effect is noticeable as the proportion of gluten in the dough is increased.

Staling is perceptible in several aspects of the organoleptic qualities of bread. The crust loses its brilliance and becomes dull, while its crispness also disappears. It may either harden or soften, depending on storage humidity, and it becomes less pleasant to the taste.

The crumb loses it supple and mellow qualities and becomes more opaque. It also becomes granular or grainy to the touch, with a tendency to crumble. It is less easily moistened, and consequently is less pleasant to chew, while the taste changes and loses much of its appeal.

The evolutionary changes that occur in staling may be influenced by many factors, including changes to dough constituents or modifications in production technology. It is generally thought that several attributes of bread flour may slow the rate of staling. For example:

- an average or above average protein level;
- a damaged starch level that is slightly higher than average, enabling better water absorption and helping the amylases to achieve a higher maltose production); and
- a slightly higher than average amylase content.

Keeping qualities may also be improved by careful attention to the following considerations during dough and bread production:

- good hydration of the flour;
- appropriate dough mixing, which ensures the formation of a good gluten network while avoiding any excessive oxidation;
- use of an additive that contains a small amount of lecithin (when needed);
- a sufficient first fermentation, or the addition of a fermented culture during mixing, which helps to develop an appropriate acidity level in the dough;
- techniques of loaf molding that avoid excessive or rough degassing of the dough,[1]
- final fermentation and baking of the formed raw dough pieces that will be adequate to produce loaves with a reasonable specific gravity (not excessive); and

- baking at an appropriate temperature, which will ensure that the loaves undergo a balanced baking (i.e., neither too hot, cool, dry, nor humid).

It should be pointed out as well that bread made by the straight dough method has advantages in terms of freshness and especially of taste, provided it receives a sufficiently long fermentation period.[2] This is also true of bread made according to production methods that include the addition of a prefermented culture—that is, natural *levain* (prefermented sponge), *levain de pâte* (naturally prefermented dough), *poolish* or Polish-style yeast sponge, *levain-levure* (sponge and dough), or prefermented yeast dough. All have prolonged shelf life and more pleasant aroma than those loaves made according to an accelerated method, such as the use of intensified mixing, in which the first fermentation period is markedly shortened.

Is it possible to interrupt the staling of bread? This is a difficult thing to do, but the temperature at which the bread is stored may be of great help. Above 30°C (86°F), the gradual change that leads to staling is relatively slow. If one keeps the bread at 60°C (140°F), staling itself will be arrested, but at 30°C (86°F) and above, in combination with the necessary humid environment, microorganisms begin to develop or threaten to do so, thus eliminating this solution.

The baker (and the consumer) should know as well that staling is accelerated when the product is stored at relatively low temperatures, especially just above the freezing point (0°C, or 32°F). However, when the product is brought to a frozen point of –20 to –30°C (–4 to –22°F) through rapid supercooling, staling is slowed or even stopped, allowing the bread to be stored in a "fresh" state.

Theoretically, when it has been frozen under commonly known and accepted conditions, it should be possible to keep bread fresh as long as desired. On a practical level, that is not true at all. In the case of French bread, baguettes and long loaves that have been frozen in the fresh state and securely wrapped to protect them from the drying effects of air may be kept for 3 to 4 days maximum under proper conditions. Beyond that, the crust tends to flake or scale off and the quality of the bread suffers accordingly. Round, compact loaves and those that have a higher fat level may be kept for 7 to 10 days in an excellent state of freshness, again provided that they are well protected from the air.

In the case of home freezers, which are unable to maintain the lower temperatures of commercial units, the storage period will be more limited: 2 to 3 days for long loaves, and 5 to 6 days for short loaves. Bread must be kept in airtight bags in this instance as well.

Bread must be thawed carefully under both home and commercial conditions:

- either at 200 to 220°C (392 to 428°F) and conditions of high humidity for 4 to 5 minutes (After this brief oven heating stage, the loaves must be set aside until defrosting of the center has taken place.)
- or at room temperature

In both instances the loaves must be protected from the drying effects of air currents.

Bread that is to be frozen must be absolutely fresh and should be placed in the freezing chamber when the cooling stage is completed. Loaves should be of a reasonable size and volume, that is, slightly less than average. This is because loaves that have excessive volume (as a result of overmixing and oxidation) generally have a thin crust, and they are thus very fragile. As a result, they are less able to resist the successive thermal shocks they experience: cooling after baking, freezing, and finally defrosting. Excessive volume may have the end result of abnormal crazing or cracking of the crust, followed by crust scaling and excessive and unpleasant crumbing.

Under ideal conditions, however, it is possible to store fresh bread through proper freezing without affecting its quality. This storage period is relatively limited for French bread, but if freezing is done correctly and the bread is fresh and of high quality, it will be just as pleasing and appetizing upon defrosting, without any staling or spoilage.

It should be noted that additives termed "anti-staling agents" are used in the production of pan breads (i.e., tinned bread). In particular, these are emulsifiers of the mono- and diglyceride types, such as calcium stearoyl-lactylate or sodium stearoyl-lactylate.

Although the use of these additives for the production of standard bread is forbidden in France, I had the occasion to experiment with them in Brazil. (The formulation of traditional Brazilian bread is quite close to that of French bread.) I found that the effect of such additives on such lean formulas was very insignificant, and furthermore that their presence had a negative effect on the taste of the crumb. In the end, it was found that a better anti-staling effect and better keeping qualities were achieved by adding a fermented culture during dough mixing. This lowered the pH of the dough mixture and noticeably improved the taste of the resulting bread.

This procedure is similar to the one used by Japanese bakers in the production of sandwich-type pan bread (tinned bread). They use a traditional sponge—a yeast-based culture that includes 70% of the flour in the formula. This allows them to substantially lower the dough pH, to retard the staling of their product, and also to improve the flavor of the bread.

CONSUMPTION OF STALE BREAD

Messrs. A. Guilbot and B. Godon, researchers at the Biochemical and Technology Laboratory of the National Institute for Agronomy Research in Nantes, discussed the problem of stale bread in a study which appeared in the *Cahiers Nutrition et Diététique*, XIX-3–1984. They noted that although the staling of French bread occurs rather rapidly, this does not seem to have had any great influence on the decline in bread consumption. This latter phenomenon appears to them to be linked to the purchasing habits of the French consumer.

In urban areas, when bread is not purchased on an every-other-day basis, it is purchased daily and for the most part is bought as long loaves. Even though this type of bread changes very rapidly, it hardly has time to dry out and become stale unless some of it is left at the end of the day. In rural areas, food shopping is done less frequently, although the custom is not as common as in the past because of the automobile. As a result, country dwellers traditionally eat more large loaves, which keep better than the long types.

It may happen that from time to time consumers are obliged to take measures to avoid having stale bread on hand. Bread is customarily reheated (to temporarily reverse staling), used in the form of bread soup, as French toast, in bread pudding, as grated bread crumbs, as ordinary toast, or even dipped into soup. These customary uses are becoming less common, however. In urban areas, stale bread may be eaten as toast or as bread crumbs, but it is often simply thrown away. In the country, it may still be used as noted above.

In both cases, stale bread may be used as animal feed. Guilbot and Godon indicate in their research report that 45% of all stale bread is used as animal feed, including 50% of stale bread from country sources, compared with 26% percent from Paris. On the whole, only 23% of stale bread is used as human food.

Unfortunately, the largest part of bread currently produced in France is made from overmixed doughs. These doughs yield bread that is often rather tasteless, and after staling it is even more inferior in this respect.

More and more often, country residents are using either the produce chilling compartment of the refrigerator, or even the freezer, to store part of the bread. They thus avoid having stale bread, which tends to diminish the overall importance of the problem of waste.[3]

SHELF LIFE AND TASTE OF INDUSTRIALLY PRODUCED PACKAGED BREADS

It is impossible to ignore industrially produced packaged breads whenever the storage and keeping qualities of bakery goods are discussed. In France, these products are even offered to the consumer from the shelves of artisan bakeries. However, they are much more commonly sold from the shelves of the whole range of grocery stores, from the smallest neighborhood grocery to the largest hypermarket.[4]

Maintaining the shelf life of bread involves not only keeping it fresh, but also preventing its spoilage. This is important for specialty breads, and more particularly, for soft sandwich breads (which are baked in an open loaf pan) and Pullman-type breads (which are baked in a covered tin). Shelf life of industrially packaged products varies greatly, ranging from 24 to 48 hours to 12 to 15 days and even as much as 8 to 10 weeks, and sometimes longer.[5]

In artisan bakery shops, specialty breads baked the same day or the day before (including soft sandwich breads made on site) are sold as whole loaves or sliced on demand while the client waits. Industrially produced medium shelf life packaged soft sliced sandwich breads are also found in such shops in France, but less and less frequently.

The specialty breads sold in the whole range of grocery stores are primarily prepackaged and sliced medium shelf life products, with a higher level of additives than is found in baked goods made in artisan bakeries. The medium and long shelf life sandwich breads sold in these stores are also presliced and prepackaged.

There are significant differences in the levels of preservative treatment applied to sliced and prepackaged specialty or soft breads that are sold fresh (with a shelf life of 24 or 48 hours), medium shelf life goods of the same general type, and long shelf life sandwich breads.

What are the effects of these different levels of treatment on taste? First of all, slicing of bread, even when it is fairly fresh, tends to release some of the volatile aroma of the crumb. Even though this amount is relatively small, preslicing bread causes damage to the taste. This is especially true of delicately flavored breads, such as those made from type 55 flour.

If the bread were not then immediately placed in an impermeable package with a very small headspace, the damage from slicing would be minimal. However, when bread is kept in a confined atmosphere for several hours, it undergoes real damage. The crust softens, its taste changes, and if part of it is slightly scorched—as often happens—the whole of the taste and odor of the bread will become atypical and often disagreeable. *It is no exaggeration to state that presliced breads kept under confined headspace conditions will undergo inevitable damage to the taste and aroma complex.*

Furthermore, a preservative (calcium propionate) is added to the dough of medium shelf life specialty and sandwich breads in order to inhibit the growth of molds for up to 12 to 15 days. This mold-inhibiting agent does have the unpleasant side effect of giving the breads a taste and an aroma that are a mixture of slightly putrid and acidic odors. These odors become concentrated in the enclosed headspace formed by the airtight bag, and when it is opened the smell is thoroughly disagreeable.

Long shelf life breads actually undergo sterilization in the course of their processing. After slicing, they are enclosed in heat-resistant airtight packaging and are then passed through an infrared tunnel oven heated to 160°C (320°F) for 8 minutes. After sterilization and cooling, they are placed in an outer protective package. Provided that the packaging remains airtight, they are protected from spoilage for weeks on end. However, this special treatment does not prevent them from drying and staling, and after about 10 days they are unfit for consumption unless they have been toasted.

Thus, in spite of all the operations after slicing that are intended to produce a more or less extended shelf life—tempering, packaging, the use of mold inhibitors, and or sterilization—it must be concluded that these measures lead to serious and irreparable damage to the quality and originality of bread taste.

Such considerations lead me to advise the consumer who loves good bread—whatever the type—to try to find good-quality specialty or sandwich breads, to buy them fresh, and if possible to have them sliced only at the time of sale. It also seems to me that a nearby neighborhood bakery would be most suitable, provided that the quality of its products is good enough.

TYPES OF BREAD SPOILAGE

Although spoilage is relatively rare with traditional breads, it may appear in the form of molds or unexpectedly as "ropy bread."

Molds

Molds are the most frequently encountered type of spoilage. However, since they do not appear until after several days of storage, they pose no danger to traditional French breads, which are intended for immediate consumption. As we have seen, molds may develop in packaged sliced specialty and soft sandwich breads after 4 or 5 days, with the likelihood increasing in relation to the moisture content of the crumb.

Medium shelf life breads—with a shelf life of 12 to 15 days—generally use two mold inhibiting agents: calcium propionate and propionic acid. Both are chemical products with fungicidal properties and may be added during the mixing stage up to the legal maximum of 0.5% based on the weight of the finished, ready-to-eat product. The addition of these products prolongs the shelf life of the treated bread as previously noted, but requires the baker to substantially increase the amount of yeast used. As we have already discussed, the baker uses sterilization to achieve long-term shelf life for specialty products. In both cases (i.e., medium and long life sliced packaged goods), these products are offered to the consumer with a sale date limit indicated on the label.

Ropy Bread

Ropy bread is caused by a bacterium (*Bacillus mesentericus*) commonly found in soil that infects bread either through flour or, less frequently, through water or contaminated yeast. The microorganism is thus found in all doughs, and the spores are able to survive the heat of baking. Under conditions favorable to the *bacillus*, a **bread infection** may begin to develop in the central portion of the crumb within 5 to 6 hours after baking and may produce a serious spoilage problem. As the bread becomes stringy or ropy, the crumb takes on a light gray tint and becomes sticky. An extremely disagreeable putrid odor and taste develop and at this stage, the crumb forms mucous strings. The bread is no longer edible but is not a health hazard.

This type of spoilage is very rare and is seen only under warm and humid conditions.[6] The risks of occurrence are greatest between 30 and 40°C (86 to

104°F). Fortunately, it is practically never encountered in breads made from acid doughs, such as those made from *levain* starters or by the *levain de pâte* method. Rope is generally limited to yeast-leavened products.

The dangers of rope infection are increased for yeast-leavened breads when

- the dough has a high water content;
- the loaves are inadequately baked and overly humid;
- tempering and cooling take place too slowly or inadequately; or
- the bread has a high proportion of crumb to crust, as with large round loaves, breads baked in tins (or pans), soft sandwich breads, and *pain biscotte*.

To avoid this serious problem, it is wise to take precautions during the critical warm and humid periods.

- Choose the cleanest and purest flours.
- Keep the bakery clean.
- Mix doughs that are rather firm.
- Encourage a vigorous primary fermentation.
- Bake the loaves thoroughly.
- Ensure rapid tempering and cooling of the loaves.

If the problem appears in spite of these measures, the baker must combat it by acidifying the dough with one of the following products: (The usage level is the maximum permitted in France for 100 kg flour.)

- pure lactic acid (300 g)
- calcium phosphate (700 g)
- pure acetic acid (200 g)
- food grade vinegar (2 L)
- calcium acetate—free of copper and zinc (430 g)

The bakehouse or bakery must also be thoroughly disinfected at the same time.

- All utensils and equipment must be washed with acidified water, i.e., treated with 200 g vinegar or 20 g acetic acid per liter of water.
- The floor must be washed with a strong mixture of water and bleach.
- The walls and ceiling must be thoroughly disinfected and whitewashed.

I should again state that although this very rare type of spoilage is spectacular, the situation is not dangerous since there is no risk that the bread will be consumed. The problem is simply a result of the environmental conditions—heat and humidity—experienced by infected loaves after baking.

NOTES

1. In regard to the effects of dough molding, it is very interesting to note that dough pieces used for the production of rustic wheat breads, which do not undergo the physical stress of dough molding, produce bread that has a shelf life at least 50% greater than baguettes made from the same dough, but which undergo dough molding stress (*see* Exhibit 10-10).

2. Long fermentation straight doughs exhibit the formation of the same types of organic acids found in prefermented doughs. It is these acids that help improve the keeping qualities.

3. There is no doubt that a large number of consumers use refrigerators for bread storage, in France as well as in North America. This practice should be discouraged, since it actually accelerates bread staling, as noted in the preceding text. Staling of bread and what to do with commercially produced stale "returns" is also a serious problem in the U.S.

4. In the United States, as opposed to the practice noted by Calvel, the sale of commercially produced prepackaged breads in artisan bakeries is almost completely nonexistent.

5. Some specialty soft breads, such as military ration buns used by the U.S. military, have a shelf life of at least 2 years under normal storage.

6. While this was doubtless true when the original French edition of this work was published in 1990, subsequent developments have caused it to become much more common, both in Europe and North America. Wet harvest conditions over several years, in combination with an emerging preference for "all-natural" breads among North American consumers, have contributed to the reemergence of ropy bread as a problem. The situation is even worse in much of Eastern Europe because of poor harvest conditions and unsanitary grain storage, milling, and bread production practices.

PART IV

Traditional and Specialty Bread Production

Basic French Bread

CHAPTER 10

- Breadmaking with *levain* and with *levain de pâte*.
- Yeast-raised French bread (*pain courant*).
- Rustic (country-style) bread with pure-wheat flour.

Having discussed at length the principal factors basic to the taste of bread, I would like to provide formulas and procedures for these same methods. Good-tasting, high-quality, distinctive breads can always be produced by using these formulas and procedures.

BREADMAKING WITH *LEVAIN* AND WITH *LEVAIN DE PÂTE*

In the most simple terms, this is a bread production method is based on inoculation or seeding of the dough with a natural leavening culture or *levain*. This procedure is carried out without the addition of industrially prepared "natural" baker's yeast—at least in principle.

Since its first appearance about 3000 years before the Christian era, this natural *levain* or sponge system has consisted of a wheat or rye dough leavened with a mixed culture of ambient yeasts and lactobacilli. Whenever and wherever bread has been made on a daily basis since that far-distant time, the baker has cultivated and maintained these mixed cultures, which were then used to build *levain* sponges. These sponges were used in turn to "seed" or inoculate the doughs for each day's production. With the passage of time, it sometimes happened that the culture became inactive or contaminated. It was customary in such cases for the baker to get a small amount of dough from a neighboring bakery. He could then renew the base cultures and continue as before.

However, it is entirely possible to "build" an entirely new culture within 3 days, provided one has some time available. Recipes for this purpose are often quite amusing, including cultures based on grape juice, potatoes, raisins, yogurt, honey, and so on.

In fact, building a culture is simply a matter of seeding or inoculating bread flour. I simply use the proper type of bread flour, and the results have always met my expectations. Table 10–1 provides a base formula that allows the baker to obtain an excellent *levain* or natural leavening culture quickly and simply, beginning with a mixture of 50% bread flour and 50% rye flour.[1]

The Food Laboratory of the National Institute of Agronomic Research in Dijon analyzed one of these cultures, with the following results for 1 g of culture:

- Yeasts: 20 million yeast cells without any *Saccharomyces cerevisiae* (the latter is common baker's yeast)
- Lactobacilli: 1.03 billion cells

These figures are comparable to those given by the laboratory of the Western Regional Research Laboratory of the University of California at Berkeley (near San Francisco). The analyses of this laboratory showed:

- Yeasts: from 15 to 28 million cells
- Lactobacilli: from 600 million to 2 billion cells

A suitable base culture may be produced from the 50% wheat flour and 50% whole rye flour mixture shown in Table 10–1, hydrated with 50% water by weight.[2] (See the Table 10–1 and the accompanying note.) To this are added 3 g salt and 3 g malt flour or malt extract, and the whole is then mixed adequately. The malt is added to increase the amylolytic power of the flour. Salt is added to protect the dough against the action of proteolytic action that might possibly weaken the gluten during the very earliest stages of dough fermentation. This proteolytic effect might otherwise damage the dough by softening it excessively, since this first fermentation stage may last for more than 20 hours. For this same reason, salt also fulfills an important role in the renewal or feeding of cultures that are ultimately to be used in the building of a naturally fermented sponge or *levain*.

During rest periods, the dough is kept at around 27°C (81°F) and protected from dehydration. As the

Table 10-1 Makeup of a Naturally Fermented Sponge from a Mixture of Wheat Flour and Rye Flour

Ambient Temperature	Duration of Maturation (time)	Proof Level Attained	Cultivated "Seed" Dough Addition	Flour	Water*	Salt	Malt
27° C (81° F)	0 h	0		600 g (300 g + 300 g)	360 g (12.698 oz)	3 g	3 g (0.1058 oz)
27° C (81° F)	22 h	2	300 g (10.58 oz)	300 g (10.58 oz)	180 g (6.349 oz)	1 g	2 g (0.0705 oz)
27° C (81° F)	7 h	3.2	300 g (10.58 oz)	300 g (10.58 oz)	180 g (6.349 oz)	1 g	
27° C (81° F)	7 h	3.5	300 g (10.58 oz)	300 g (10.58 oz)	180 g (6.349 oz)	1 g	
27° C (81° F)	6 h	4.3	300 g (10.58 oz)	300 g (10.58 oz)	180 g (6.349 oz)	1 g	
27° C (81° F)	6 h	4.3	300 g (10.58 oz)	300 g (10.58 oz)	180 g (6.349 oz)	1 g	
27° C (81° F)	6 h	4.1	300 g (10.58 oz)				

*Professor Calvel himself adjusted the hydration rates for North American flours from 50% to an average of 58%, based on work that he had done with North American flours both at Kansas State University and at the Culinary Institute of America. Some flours required a hydration rate up to 62%. In all cases, the object is to produce a rather firm dough. (Personal communication from Professor Raymond Calvel to James MacGuire, 1999.)

---

culture gradually develops, the amount of fermentation activity of the yeasts and lactobacilli improves, and the volume level of dough rise tends to increase. The figure given in the third column of the Table 10–1 indicates the maximum level of rise, beyond which the dough does not "move," shows signs of fatigue, and has a tendency to become porous. In fact, the level of dough rise after 5 to 6 hours is less than level 4 in the graduated cylinder (shown with the *levain* discussion below), from which it may be concluded that such a culture is no longer active.

The daily bread of Professor Calvel's youth in the Haut Languedoc was pure *levain*. This was replaced by the hybrid *levain de pâte* which in turn was replaced by the yeast-based recipes in the following sections. Despite a celebrated *levain* specialist or two, most bakers who produce *pain au levain* do so as a sideline to their yeast-raised *pain courant*. These loaves, more delicate in flavor and texture than San Francisco sourdoughs, deserve more attention.

At this point it should be noted that the volume level of dough rise produced by the culture is not the only criterion to bring into consideration. The acidity or pH of the dough must also reach a proper level, which is of prime importance to the taste of bread. Thus, a culture will be suitable for "seeding" a *levain* when it rises to a level of 4 to 4.2 in 6 hours, and has a pH ranging from 4.4 to 4.6. Such a culture is normally capable of producing good-quality naturally leavened bread.

Exhibit 10–1 provides—at least in principle—the steps necessary to the building of a naturally leavened sponge that will allow the baker to obtain similarly good results.

In only 54 hours—that is, 2 days and 6 hours—a natural culture of this type can be used for practical operations. With the last culture in the series, the baker may prepare a naturally fermented sponge, termed the primary culture *levain*, which is used to inoculate a dough that will produce bread from a natural sponge fermentation process. It is certainly true that this gradual building does not always have the regularity of a metronome, but it is rare not to succeed in obtaining excellent results.

Breadmaking with a natural *levain* requires the ongoing maintenance of the fermented culture of yeasts and lactobacilli that are the basis of this type of fermentation. On a daily production basis, the sponge is prepared during the interval following the end of the day's production and the mixing of the first batch of the succeeding day. Five or six hours before mixing the dough for the first oven loading, the sponge is inoculated or seeded with a dough piece that is known as the "chef" in France. This dough may be inoculated either once or several times.

When the dough is inoculated one time, the breadmaking procedure is called "working from a single *levain*." When the dough is inoculated twice, it is known as "working from two *levains*."[3] In this case, the first culture is called the "renewer" or "refreshing" culture. In the past, many bakers used a three-stage culture or "three *levain* culture" method in the makeup of such natural sponge leaveners.

Today, the equipping of bakeshops with walk-in coolers or refrigerators simplifies the preservation and storage of the chef and allows the use of one or two cultures in breadmaking. However, the storage temperature should be kept at 10°C (50°F) or slightly higher in order to preserve intact the flora that make

Exhibit 10–1 Steps to Build a Naturally Leavened Sponge

Refresher Culture (Rafraîchi)

Chef (starter)	520 g (18.34 oz)	
Flour	675 g (23.81 oz)	
Water	405 g (14.285 oz)*	
TOTAL	1,600 g (56.437 oz.) of hydrated dough	60% hydration
Low speed mixing	10 min	(mix until smooth dough stage)
Dough temperature	25–26° C / 77–79° F	depends on ambient conditions
Proof time	5 to 6 h	rise to 3.5 times starting volume

Natural Fermented Culture (Levain)

Rafraichi (refresher culture)	1600 g (56.437 oz)	
Flour	1760 g (62.08 oz)	
Type 85 Rye Flour	155 g (5.467 oz)	
Water	1150 g (40.564 oz)*	
TOTAL	4,665 g (164.55 oz) of hydrated dough	60% hydration
Low speed mixing	10 min	(mix until smooth dough stage)
Dough temperature	25–26° C / 77–79° F	
Proof time	4 to 6 h	rise to 3.5 times starting volume

Pain au Levain (bread made from a naturally fermented sponge)
Formula
(For each 25 parts of naturally fermented sponge allow 100 parts of flour)

				Total Build Ingredients	*Dough Makeup*
Flour	19,000 g (670.194 oz)*			2435 g (85.89 oz)[†]	16,565 g (584.30 oz)
Type 85 Rye Flour	1,000 g (35.27 oz)*			155 g (5.467 oz)	845 g (29.805 oz)
Total flour	20,000 g or 20 kg (705.467 oz)		100%	2,590 g (91.358 oz)[†]	17,410 g* or 17.41 kg (614.109 oz)
Water	12,800 g (451.499 oz)[†]		64%	1,555 g (54.85 oz)	11,245 g (396.649 oz)
Chef (starter)				520 g (18.34 oz) (325 g [11.46 oz] flour + 195 g [6.878 oz] water)[†]	
Salt	350 g (12.35 oz)		1.8%		360 g (12.698 oz)
Yeast (occasionally)[1]	40 g (1.41 oz)		0.2%		40 g (1.41 oz)
Natural *levain* (sponge)			—	4,665 g[†]	4,665 g (164.55 oz)
TOTAL DOUGH WEIGHT					33,680 g or 33.68 kg (1188.007 oz)
Low speed mixing					5 min
Autolysis[2]					30 min
Low speed mixing					12 min
Dough temperature					25–26° C / 77–78.8° F
Primary fermentation					50 min
Division & rounding (put aside chef for next day's production)					10 min
Bench time/rest stage					20–30 min
Molding					10 min
Secondary fermentation					About 4 h
Proof level					3.5 to 4 times
Baking (230° C / 446° F)[3]					30–40 min
TOTAL MAKEUP AND BAKING TIME INVOLVED					7 h 10 min

* There was an error in the original French edition, for which Calvel provided this correction for the English language edition.
[†] The *chef* appears to have been omitted from calculations for the original French version.

[1] When a small amount of baker's yeast is to be added, the use must be 0.2% at most, to avoid adulteration of the characteristics of naturally leavened bread (*pain au levain*). To preserve the purity of the primary culture, the baker may draw off the chef, which will be used for the production of future sponges, after 6 or 7 min of mixing and then add the yeast. In that way the chef will remain a culture that has been seeded only with wild yeasts.

[2] Dough autolysis refers to a rest period that occurs after 5 min of mixing a fraction of the flour and part of the water, excluding the remaining ingredients. This rest period improves the links between starch, gluten, and water and notably improves the extensibility of the dough. As a result, when mixing is restarted, the dough forms a mass and reaches a smooth state more quickly. Without autolysis, these processes would be slowed by the acidity of the natural *levain*. Autolysis also facilitates molding of the unbaked loaves and produces bread with more volume, a better cell structure, and more supple crumb. Although the use of autolysis is advantageous in the production of several types of bread, including common French bread, white pan sandwich bread, and sweet yeast doughs, it is especially valuable in the production of natural *levain* leavened breads. (See discussion of autolosis in Chapter 3.)

[3] Baking naturally leavened breads requires careful attention. For a number of reasons, the loaves must be baked at a lower temperature than those produced with baker's yeast:
• Oven spring is slower with naturally leavened breads, and it is important to get maximum rise before crust formation.
• Precisely because the expansion of the dough piece occurs more slowly, naturally leavened loaves are generally somewhat compact in form, being either rounded or slightly elongated. Because volume development occurs more slowly, it must not be inhibited by too high a baking temperature.
• In the interest of flavor, it is important that the loaf have a good crust with a warm, rich tint, but without being excessively colored. (An overbaked crust is really very unattractive.) To produce a good crust, the baker must anticipate a relatively long bake time, and excessive oven temperature must be avoided.

up the natural *levain* or "sourdough." At a temperature lower than 8 to 10°C, part of the flora is damaged, and the bread loses some of its distinctive characteristics. That is not to say that fermentation is inhibited: the chef and the sponges rise correctly, but the resulting loaves do not have the distinctive aroma of bread made with a natural *levain*.

It is regrettable that the temptation to preserve doughs at too low a temperature or to freeze them is so great. As a consequence, it is rare today to bite into a *levain*-fermented bread that retains the original and distinctive taste characteristics. Two examples for the proper storage of master cultures are shown in Figure 10-1.

Another subject that should be discussed is the addition of baker's yeast while breadmaking with a *levain*. This is done quite often, but when the baker is making and selling bread that is supposed to be from a natural *levain*, the addition of yeast is contrary to established custom. In any case, when the baker cultivates natural *levains* without any addition of baker's yeast, and the yeast is only added at the rate of (0.1% to 0.2%) at the end of mixing (that is, from 1.5 to 3.0 g per liter of water), the taste of the resulting bread is not much affected and the baker gains precious time during the second fermentation.[4]

It certainly is out of the question for a baker who is making "naturally fermented" bread to use more than 3 g yeast per liter of water (more than 0.2%), and then to continue to sell the resulting bread as being produced from a natural *levain*. Above that 0.2% level, the bread is no longer made from a natural sponge but from a prefermented yeast dough or *levain de pâte*.

This latter type of bread was commonly made during the 1920–1930 period, with the addition of around 0.5% yeast. When the technique was performed correctly, the baker produced bread of excellent quality that was higher in volume, but with a less pronounced *levain* taste and a more limited shelf life.

Formulations for breads made with a natural *levain* and with a prefermented dough or *levain de pâte* follow.

Formula for Bread Made with Natural *Levain*

The formula in Exhibit 10-1 for the production of bread from a natural *levain* has been shown to give very regular and extremely good results, both in France and elsewhere. It includes two successive cultures, a refresher culture and a *levain* or fermented sponge, and thus is termed "work from two *levains*." The first refresher culture is built from the chef, a small amount of sponge or of unbaked dough from the previous day's baking.

When the *chef* is used to make up the refresher culture, the volume of the refresher should reach at least 3.5 times the beginning volume (Figure 10-2). During the makeup of the second sponge, the *levain*, the addition of 5% type 55 rye flour is advisable in order to improve the taste and keeping quality of the bread.

Formula for Bread Made with *Levain de Pâte*

The method used in this case involves the building of only a single *levain* from the chef, which was taken from a batch made the previous day and kept cool (10°C). The *levain* is made up 5 or 6 hours before mixing the new dough and is prepared as shown in Exhibit 10-2.

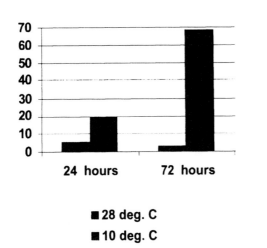

Figure 10-1 Extended Refrigerated Storage of "Chef" or Master Culture

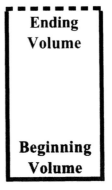

Figure 10-2 Schematic View of the Rise of Natural Sourdough.

Exhibit 10–2 Levain de Pâte

Makeup of the leavening sponge (levain)
Characteristics of the master culture (chef): weight 2.5 kg
Proofing time: 10 h at 15–18° C

Culture (chef)	715 g (25.22 oz) flour (+ starter) 1.515 kg + 985 g (53.439 oz + 34.88 oz)
Flour	2860 g (100.88 oz)
Water	1715 g (60.493 oz) total weight 5290 g or 5.29 kg (186.596 oz)
Mixing: 1st speed	10 min
Dough temperature	25–26° C / 77–79° F
Proof time	5–6 h @ 24° C / 75° F

Levain loaves have better oven spring when shaped in round or slightly oblong loaves.

Makeup of the dough
Allow 30 parts of leavening culture for each 100 parts of flour

Formula			Portion from Sponge	Final Dough Makeup
Flour	20 kg (705.467 oz)		2860 g or 2.86 kg (100.88 oz)	17.14 kg (604.585 oz)
Salt*	400 g (14.109 oz)	2%		400 g (14.109 oz)
Yeast†	200 g (7.055 oz)	1%		200 g (7.055 oz)
Water	12,800 g (451.5 oz)	64%	1715 g (60.49 oz)	11,085 g (391.005 oz)
Ascorbic acid	(400 mg)	20 ppm		400 mg
Chef Weight			715 g (25.22 oz)	
Sponge Weight			5290 g (186.596 oz)	5290 g (186.596 oz)
TOTAL DOUGH WEIGHT				34,115 g or 34.115 kg (1203.351 oz)
Mixing: 1st speed				5 min
Autolysis				30 min
Mixing: 2nd speed				10 min
Dough temperature				25° C / 77° F
1st fermentation				45 min
Dividing + rest				35 min
Tourne = shaping				
2nd fermentation	27° C / 81° F			110 min
Baking	230° C / 446° F			25–40 min
TOTAL TIME				4 h 45 min

*Technical note: in this and following recipes Calvel has listed a salt content of 2% by flour weight, whereas in Chapter 2 he advocated a 1.8% salt level. In personal communication Calvel has noted that the additional 0.2% salt could be useful in providing flavor if the dough is overkneaded, and therefore lacking flavor. If the dough is properly kneaded (developed) the 1.8% salt remains a good idea.

†Delay the addition of the yeast until the 6th minute of mixing time, being sure to remove the *chef* for the next day's production before adding the yeast.

Exhibit 10–3 Basic French Bread: Straight Dough Process with Traditional Mixing Method (1st speed for all mixing)[1]

Formula

Wheat flour	20 kg	100%
Water (after 15 minute autolysis)	13.4 l	67%
Yeast	250 g (8.82 oz)	1.25%
Malt extract	40 g (1.41 oz)	0.2%
Salt (after autolysis)	400 g (14.11 oz)	2%
TOTAL WEIGHT OF DOUGH	34.09 kg	

Procedures

Mixing: 1st speed	3 min
Autolysis	15 min
Mixing: 1st speed	12 min
Dough temperature	24° C
(from mixer)	75° F
1st Fermentation	150 min
Punch down after	80 min
Dividing, rounding & rest	35 min
Molding	as needed
2nd fermentation	75 min
Baking 240° C / 464° F	30 min
TOTAL TIME	6 h 40 min

[1]This recipe seeks to recreate the older style of bread that Professor Calvel so elegantly describes in Chapter 4. Mixing is done at slow speed only, and the first fermentation must make up for the dough's lack of strength. The mixing times with other types of mixers would vary as follows: spiral mixer (3 min as noted + 4 min); Artofex mixer (3 min + 10 min); planetary mixer (3 min plus 8 min).

The type of bread in the following section, once called *pain de fantaisie* but now called *pain courant* or *pain ordinaire*, is known around the world as *French bread*. Most often made in baguette form, it is by far the largest selling bread type.

The different formulae/recipes may therefore be considered variations on a theme, producing subtle differences which help to define each baker's style. No bakery would ever produce all of the recipes, and small bakeries often produce only one. Those who do feature more than one might choose two distinct styles, for example, the prefermented dough for regular French bread (Exhibit 10–9), and the tradionally-mixed straight dough method to satisfy the increasing demand for "retro-baguettes" (Exhibit 10–3).

YEAST-RAISED FRENCH BREAD (*PAIN COURANT*)

For the most part, bread is made today by leavening it with baker's yeast, a natural biological substance. As we have already noted, there are several methods, including

- the straight dough method, which is the most widely used, and
- methods based on a fermented culture
  —the *poolish* or Polish-style sponge method
  —the sponge and dough method
  —the added preformented dough method.

The Straight Dough Method

This method produces excellent quality bread, provided that mixing is adequate and appropriate. It also requires a relatively long first fermentation to achieve natural dough maturation, an absolutely essential element in the production of quality bread. However, with the mechanized methods used today, this is often difficult to do. Breadmaking schedules developed to facilitate the mechanical operations that the dough undergoes tend to reduce or effectively eliminate the important first fermentation.[5] The natural maturation produced by this fermentation is replaced by an artificial one achieved by the addition of additives and the hyperoxidation of the dough through intensive mixing.

This artificial maturation has the effect of suppressing the formation of the important organic acids that contribute to the creation of bread taste and also to the improvement of its keeping qualities. The result of these artificial practices is irreversible damage to the end product, especially to the taste and aroma of both dough and bread.

Use of the formula and procedures outlined in Exhibit 10–3 results in the production of bread that has a creamy crumb, excellent flavor, and very good quality overall. There may be some difficulty in using production equipment with this formula, especially a semiautomatic divider. Dividing may be difficult, and the dough pieces somewhat irregular in weight. After a reasonable rest period for the dough pieces, however, they may be passed through the divider, and because of the beneficial effect of autolysis, it is possible to roll them to the proper length without too much difficulty. When the dough is divided and formed by hand, it produces loaves of exceptional quality (Figure 10–3).

The formula and procedures given in Exhibit 10–4 also produce excellent quality bread. The dough will give fewer problems with a semiautomatic divider, and the dough pieces may be properly formed with an automatic molder after an appropriate rest period. The loaves are noticeably higher in volume than those produced by conventional mixing, and they have a light cream-colored crumb and an agreeable taste. They also have pleasant mouthfeel and good keeping qualities.

The Straight Dough Method with Intensive Mixing

As I stated previously, the only reason to even mention the straight dough method with intensive mixing (Exhibit 10–5) is to point out that the resulting product is no longer bread. It is so debased and denatured

The baguette has become a rather hackneyed symbol of French life, but it does not have a long history. Following the First World War, the technology was at last in place to produce light and delicately flavored loaves with a crispy crust. Mixing machines, stronger flours, yeast-based recipes, steam injection ovens, etc., all contributed to this. Consumer appreciation of the thinner, more delicate crust led to the introduction of longer loaves, with a higher crust to crumb ratio. Their higher price and the fact that they were bought and eaten on the same day caused them to be called *pains de fantaisie*, the word *fantaisie* meaning both extravagance and whimsy. By contrast, traditional sourdough or *levain de pâte* loaves kept for many days, and frugal housewives never served them fresh.

Although the baguette is the most popular French loaf worldwide, the doughs in Exhibit 10–2 through Exhibit 10–9 can be shaped into various other sizes. See Table 7–1 and Figure 7–12.

Exhibit 10–4 Basic French Bread: Straight Dough Method with Improved Mixing

Formula

Wheat flour	20 kg (705.467 oz)	100%
Water	13.4 l (472.66 oz)	67%
(Add items below after autolysis)		
Yeast	300 g (10.58 oz)	1.5%
Malt extract	40 g (1.41 oz)	0.2%
Lecithin*	15 g (0.529 oz)	0.075%
Ascorbic Acid	(400 mg)	(20 ppm)
Salt	400 g (14.11 oz)	2%
TOTAL WEIGHT OF DOUGH	34.15 kg (1204.59 oz or 75.29 lbs)	

Procedures

Mixing: 1st speed	less than 5 min
Autolysis	15 min
Mixing: 2nd speed	10 min
Dough temperature	24° C
(from mixer)	75° F
1st fermentation	60 min
Punch down after	(20 min)
Dividing, rounding & rest	30 min
Molding	as needed
2nd fermentation	90 min
Baking 240° C / 464° F	30 min
TOTAL TIME	4 h

*Lecithin is optional, and is used under conditions of large-scale mechanized production.

Figure 10–3 Traditional Baguettes.

and so different in physical characteristics and taste from real bread that it is only an imperfect caricature of it. For these reasons, I am sure that I will be pardoned for giving only the most rudimentary formula and procedure for it.

Breadmaking from a Prefermented Culture Based on Baker's Yeast

A number of breadmaking methods are based on the use of a baker's yeast culture that has been prefermented before dough mixing. These methods are better adapted to time constraints than are natural ferments, since they permit both rapidity of produc-

Exhibit 10–5 Straight Dough Method with Intensive Mixing (**not recommended**)

Formula

Flour	100%
Water (a little less)	65%
Salt (more than usual)	2.2 to 2.3%
Yeast	2 to 2.5%
Ascorbic acid (more than usual)	40 to 60 ppm

Procedures

Mixing: 1st speed	4 min
Mixing: 2nd speed	18–20 min
Dough temperature	24–25° C / 75–77° F
1st Fermentation (in mixing bowl)	5–10 min
Division, rest, molding	25–30 min
2nd fermentation	2.5 to 3 h
Baking	25 to 35 min
TOTAL TIME	4 h 39 min

General observations: to bleach, debase, and wash out the taste even more, add salt 4 to 5 min before the end of mixing.

tion and use of mechanical production equipment, while at the same time allowing the baker to maintain the bread's authentic quality and originality.

The speed of production is due to the fact that part of the alcoholic fermentation is carried out well in advance of dough makeup. It is this fermentation process that contributes priceless flavor-generating organic fatty acids to the dough and thence to the baked bread. Ascorbic acid also increases production speed by allowing the use of mechanical makeup equipment, while at the same time permitting a noticeable reduction in the time of first fermentation without any harmful effects.

Breadmaking with a *Poolish* (Polish Sponge)

The Polish sponge method was developed in Poland during the 1830s, when industrially produced baker's yeast first made its appearance. It was later adopted by the bakers of Vienna. The *poolish* method was used to make the very first breads leavened entirely with baker's yeast (Figure 10–4). In France, loaves of this type were to become widely known and were produced under the name of "Vienna breads."

This method involves the makeup of a relatively liquid yeast culture well in advance of mixing the dough proper. Both the volume and the culture fermentation time vary a great deal. The amount of water used is based on the total amount of formula water to be used in dough mixing. The sponge water may represent from one third to three fifths of total formula water—and sometimes even more. As far as length

Figure 10–4 Poolish Loaves. When more than one type of baguette is offered, the slashing method can be varied. Here they have been cut in criss-cross fashion.

of sponge fermentation is concerned, it should be a minimum of 3 hours, but may last as long as 7 to 10 hours or even longer, depending on the quantity of yeast used in the sponge and the ambient temperature. Examples of the use of a *poolish* are given in Exhibit 10–6 and Exhibit 10–7.

The fermentation time in both the first and second proof periods should certainly be adjusted slightly to take into account flour quality, dough temperature, and ambient temperatures. Thus, they may vary slightly from the production outline given.

Breadmaking by the Yeast Sponge and Dough Method

The sponge and dough method (Exhibit 10–8) is somewhat comparable to the Polish sponge method in that it involves the use of a prefermented yeast culture. At the present time in France it is seldom used in the production of traditional breads; however, it is more commonly used to produce country-style loaves and yeast-raised sweet doughs such as *brioches* and white pan sandwich breads.

The traditional way of making white pan bread in the Anglo-Saxon countries was the sponge and dough method, but it has largely fallen out of favor

Exhibit 10–6 Basic French Bread: Breadmaking with *Poolish* (using one-third of formula water, no autolysis)

Formula

	Weight		Poolish Sponge	Dough
Flour	20.00 kg (705.47 oz)	100%	4 kg (141.09 oz)(25%)	16.00 kg (564.37 oz)
Water	13.6 kg (479.72 oz)	68%	4 kg (141.09 oz)	9.6 kg (338.62 oz)
Salt	400 g (14.11 oz)	2%		400 g (14.11 oz)
Yeast	240 g (8.47 oz)	1.2%	60 g (2.17 oz)	180 g (6.35 oz)
Malt extract	50 g (1.76 oz)	0.25%		50 g (1.76 oz)
Ascorbic acid	(300 mg)	(15 ppm)		(300 mg)
SPONGE WEIGHT			8.06 kg (284.3 oz)	8.06 kg (284.3 oz)
TOTAL DOUGH WEIGHT				34.29 kg (1209.52 oz)

Procedures

Mixing: Low speed			3 min	4 min
Mixing: Second speed				10 min
Dough temperature from mixer			25° C / 77° F	24° C / 75° F
First fermentation/first proof			(4 h)	70 min
Punch down after				(30 min)
Division, rounding, rest period				30 min
Dough piece molding				
Second fermentation/final proof				80 min
Baking @ 245° C / 473° F				25–35 min
TOTAL PRODUCTION TIME				varies, 3 h 49 min

Exhibit 10–7 Basic French Bread: Breadmaking with *Poolish* (using four-fifths of formula water, no autolysis)

Formula

	Weight		Poolish Sponge	Dough
Flour	20.00 kg (705.47 oz)	100%	10.00 kg (352.73 oz)(50%)	10.00 kg (352.73 oz)
Water	13.6 kg (479.72 oz)	68%	10 l (352.73 oz)	3.6 kg (126.98 oz)
Salt	400 g (14.11 oz)	2%		400 g (14.11 oz)
Yeast	200 g (7.05 oz)	1.0%	70 g (2.47 oz)	130 g (4.59 oz)
Malt extract	60 g (2.12 oz)	0.3%		60 g (2.12 oz)
Ascorbic acid	(400 mg)	(20 ppm)		(400 mg)
SPONGE WEIGHT			20.07 kg	20.07 kg (707.94 oz)
TOTAL DOUGH WEIGHT				34.26 kg (1208.47 oz)

Procedures

Mixing: Low speed			3 min	4 min
Mixing: Second speed			3 min	10 min
Dough temperature from mixer			25° C / 77° F	24° C / 75° F
First fermentation/first proof			(7 h)	60 min
Punch down after				(25 min)
Division, rounding, rest period				30 min
Dough piece molding				
Second fermentation/final proof				90 min
Baking @ 245° C / 473° F				25–35 min
TOTAL PRODUCTION TIME				varies, 3 h 49 min

Exhibit 10–8 Basic French Bread: Breadmaking by the Sponge and Dough Method (no autolysis)

Formula

	Weight		Yeast Sponge	Dough
Flour	20.00 kg (705.47 oz)	100%	4.00 kg (141.09 oz)	16.00 kg (564.37 oz)
Water	13.6 kg (479.72 oz)	68%	2.56 kg	11.04 kg (389.42 oz)
Salt	400 g (14.11 oz)	2%		400 g (14.11 oz)
Yeast	300 g (8.47 oz)	1.5%	40 g (1.41 oz)	260 g (9.17 oz)
Malt extract	40 g (1.41 oz)	0.2%		40 g (1.41 oz)
Ascorbic acid	(400 mg)	(20 ppm)		(400 mg)
SPONGE WEIGHT			6.6 kg (232.8 oz)	6.6 kg (232.8 oz)
TOTAL DOUGH WEIGHT				34.34 kg (1211.29 oz)

Procedures

		Yeast Sponge	Dough
Mixing: Low speed		8 min	4 min
Autolysis rest period			13 min
Mixing: Second speed			8 min
Dough temperature from mixer		24° C / 77° F	25° C / 75° F
Yeast sponge fermentation		4 h	
First fermentation / proof			45 min
Division, rest period			35 min
Dough molding			
Second fermentation / final proof			90 min
Baking @ 245° C / 473° F			30 min
TOTAL PRODUCTION TIME			3 h 49 min

today. However, sponge and dough remains the current method used in Japan, where bakers produce one of the finest white pan breads in existence.

The sponge and dough method uses a culture made up in advance of mixing. The sponge that gives this method its name is rather firm in consistency and, like the more liquid *poolish*, is made up of a mixture of flour, water, and yeast, with the exclusion of salt. It is made up of a portion of the total formula flour, and the proportion used varies according to the strength of the organic acid fraction that the baker wishes to produce.

The sponge is considered to be sufficiently mature when it reaches 4 times the starting volume, generally after a minimum of 3 hours of fermentation. In French practice, once it has reached that maturity level, it is most often used in the production of a traditional bread, but it may also be used in making other types of goods.

Breadmaking with the Addition of Prefermented Dough

A prefermented culture added at the time of mixing may be a culture prepared especially for this purpose, such as a *poolish* or a sponge and dough method yeast sponge, or a piece of dough taken from a recent production batch.

This dough piece will normally have been leavened with 1.5%–2% baker's yeast based on flour weight. It will have fermented for 3 to 5 hours at ambient temperature, or it may have been kept under refrigeration for 15 to 20 hours at 4°C (39°F), or for an even longer period at a lower temperature. Since this dough piece contains salt, it will be slightly different when used as a prefermented culture, but this causes almost no difficulties whatsoever.

The flour fraction of the total flour weight contributed by this prefermented dough may vary according to the type of product being made. It is generally around 15% for traditional breads, 30% for baguettes and small Vienna breads, 30% for croissant dough, and up to 50% or more for pan sandwich bread.

Prefermented dough is the most practical method of using a culture of ferments and does not pose any particular problems. However, it is advisable to store it in plastic or metal containers to avoid surface crusting or contamination and to protect it from the air. It is likewise important to incorporate it into the new dough halfway through the mixing stage (Exhibit 10–9) to avoid excessive oxidation. There are few exceptions to this practice.

Basic French Bread

Exhibit 10-9 Basic French Bread: Breadmaking with Addition of Fermented Dough (**with** autolysis)

Formula

	Weight		Fermented Dough	Dough
Flour	20.0 kg (705.47 oz)	100%	2.8 kg (98.77 oz) (14%)	17.2 kg (606.7 oz)
Water	13.6 kg (479.72 oz)	68%	1.9 kg (67.02 oz)	11.7 kg (412.7 oz)
(After 13 minute autolysis period noted in procedure summary below)				
Salt	400 g (14.11 oz)	2%	55 g (1.94 oz)	345 g (12.17 oz)
Yeast	400 g (14.11 oz)	2%	55 g (1.94 oz)	345 g (12.17 oz)
Malt extract	40 g (1.41 oz)	0.2%		40 g (1.41 oz)
Ascorbic acid	(400 mg)	(20 ppm)		(400 mg)
SPONGE WEIGHT			4.81 kg (169.66 oz)	4.81 kg (169.66 oz)
TOTAL DOUGH WEIGHT				34.4 kg (1213.40 oz)

Procedures

Mixing: Low speed	3 min	4 min
Autolysis rest period		13 min
Mixing: Second speed	7 min	8 min
Dough temperature from mixer	24° C / 75° F	25° C / 77° F
Fermented dough rest or storage time	4 h @ 24° C or 15 h @ +4° C / 39° F	
First fermentation / proof		45 min
Division, rest period		30 min
Dough molding		
Second fermentation / final proof		90 min
Baking @ 245° C / 473° F		30 min
TOTAL PRODUCTION TIME		3 h 40 min

RUSTIC (COUNTRY-STYLE) BREAD WITH PURE-WHEAT FLOUR

I first conceived and carried out this bread production method in the spring of 1983. It is the result of all-wheat flour, without any additives, and combines a number of definite advantages. The method is simple, quick, and economical and produces a distinctive bread of exceptional quality (Exhibit 10-10).

The flour should be of good quality and free of any additives (except for cereal amylases if needed). It should contain neither legume (fava or soybean) flour nor ascorbic acid. The flour is improved by having a minimum 12- to 15-day storage period after milling during the warm season, and 20 to 25 days during colder weather.[6]

The dough is made up with an addition of prefermented dough equivalent to 12% to 18% of the total formula flour weight. The prefermented dough should be kept under ambient conditions (about 24 to 25°C / 75 to 77°F) for around 4 hours, or under refrigeration at 4°C / 39°F for 15 to 18 hours. It should be added to the new dough about 6 minutes before the end of mixing.

Mixing effectiveness will be improved by an autolysis rest period of 15 minutes minimum. Mixing at second speed should not exceed 8 minutes, with salt being added at the end of the autolysis rest period,

Exhibit 10–10 Rustic Bread from Pure Wheat Flour (no ascorbic acid or bean flour—wheat flour only)

Formula*

	Weight		Fermented Dough	Dough
Flour	20.00 kg (705.47 oz)	100%	3.00 kg (105.8 oz)	17.00 kg (599.65 oz)
Water	13.8 kg (486.77 oz)	69%	2.07 kg (73.02 oz)	11.73 kg (413.76 oz)
Salt	400 g (14.11 oz)	2%	60 g (2.12 oz)	340 g (11.99 oz)
Yeast	400 g (14.11 oz)	2%	60 g (2.12 oz)	340 g (11.99 oz)
Malt extract	40 g (1.41 oz)	0.2%		40 g (1.41 oz)
SPONGE WEIGHT			5.19 kg (183.07 oz)	5.19 kg (183.07 oz)
TOTAL DOUGH WEIGHT				34.64 kg (1221.87 oz)

Procedures

Mixing: Low speed	4 min	4 min
Autolysis rest period	18 min	
Mixing: Second speed	10 min	8 min
Dough temperature from mixer	24° C / 75° F	24° C / 75° F
Fermented dough rest or storage time	4 h @ 24° C or 15 h @ +4° C	
First fermentation/proof		45 min
Weighing, rest period in containers		25 min
Division & depositing on cloth		
Second fermentation/final proof		60 min
Baking @ 250° C oven sole, 275° vault (482° and 528° F respectively)		30 min
TOTAL PRODUCTION TIME		3 h 10 min

*The formula as given in the original French text included a small amount (400 mg) of ascorbic acid. However, the text states that is to be omitted. James MacGuire's communications with Professor Calvel in 1999 indicate that this deletion is correct and appropriate.

just before beginning the second mixing stage. The first fermentation period noted is appropriate for flour with good strength.

At the end of the first fermentation period, the dough is weighed off into containers that have been well coated with flour. Try to weigh off the dough in single pieces, placing the smoothest and most homogeneous surface of the dough piece face down in the container, without any molding or shaping whatsoever.

After a rest period of 25 minutes, gently divide the dough pieces into individual units of approximately 300 g each, either by hand using a dough scraper, or with a hydraulic divider (Figure 10–5). Deposit these individual units onto a flour-dusted cloth for a further rest / fermentation period of 45 to 60 minutes. The dough pieces may then be divided into 75 g segments with a dough knife or scraper to produce excellent small rolls.

Finally, use a baker's peel or a canvas oven loader to place the loaves into the oven, which should be lightly injected with steam. The smooth, flour-dusted faces of the rolls or loaves should be facing up and should be lightly cut with one or more blade strokes before being placed in the oven.

Bake the loaves at normal temperature. However, if it is possible to regulate the temperatures of the oven sole and oven chamber separately, it is advantageous to have the temperature of the chamber 25°C higher than that of the sole for the first 8 to 10 minutes of baking—for example, 250°C at the sole and 275°C in the chamber. This facilitates the vertical development of the loaves, which is usually desirable since they have not been molded in any way. These loaves or rolls all benefit from having a good, well-baked crust.

Figure 10–5 Dividing Rustic Loaves by Hand. This may also be accomplished with a suitably gentle dough dividing machine.

Following these few simple rules produces a bread of a rare quality: it is light with a cream-colored crumb composed of well-developed cells and has an enticing aroma. This is a very flavorful and appetizing bread that may be kept for 36 to 48 hours without any problems.

NOTES

1. A mixture of 75% medium rye flour and 25% dark rye flour should correspond closely to the type 170 rye flour in the original French text.
2. North American flours absorb more water—sometimes much more. Figures from Professor Calvel's earlier work in North America show hydration rates in the upper 50% through lower 60%.
3. To use the *levain* for recipes involving one or two "builds."
4. French law has now adopted the 0.2% maximum yeast content figure for bread to be allowed to be called *pain au levain*.
5. Well-fermented dough pieces tend to tear and cause other problems when mechanized. See Chapter 6 for further discussion.
6. White flour for all recipes benefits from the aging period noted here. Flours that are freshly milled and do not undergo this aging and natural oxidation process do not perform as well as aged flours.

Chapter 11

Specialty Breads

- Specialty breads.
- Breads for filling or topping.
- Savory and aromatic breads.

SPECIALTY BREADS

Country-Style Bread

Country-style bread[1] epitomizes the dominant aspects of the countryside and the supposed virtues that are attributed to it: naturalness, simplicity, and a rustic wholesomeness. To achieve such results,

- the flour used should be a type 55, naturally pure flour that may occasionally include a small addition of "gray" high extraction, high ash flour (farine bise), or type 130 rye flour (medium rye flour);
- the mixing stage of dough production must avoid excessive mechanical working of the dough or overoxidation;
- dough fermentation should include a very long first fermentation, or even better, the addition of a fairly large proportion of a fermented culture, such as *levain de pâte, levain-levure, poolish,* or *pâte fermentée*;
- the dough pieces should be fairly tightly formed during molding, and the second fermentation should not be overly long or excessive;
- whether the dough pieces are flour dusted or not, baking should be at a slightly reduced temperature, and steam injection may or may not be used;
- the bread must be thoroughly baked;
- the crumb should have a clear light cream tint with slight grayish undertones;
- the crumb structure should be open, with irregularly sized gas cells or alveoli;
- the odor and the taste of the bread should be tempting, pleasant, and more pronounced than "traditional" French breads, but without any noticeable acidity;
- the flavor should be pleasant and appetizing, and in addition the bread should exhibit very good keeping qualities for 2 or 3 days.

The country-style breads (Figure 11–1) found today in the larger cities should exhibit the qualities described above in order to be truly comparable to authentic *pain de campagne*, which in the past was widely produced and eaten in the majority of rural France. Unfortunately, that is seldom the case.

The baker often uses improper flour blends, exaggerating the amount of rye flour and especially of "gray," high bran content flours in order to better simulate the "authentic country" character. The doughs are overly firm, made up by the straight dough method according to a production schedule that is much too

---

The *pain de campagne* revival phenomenon began in Paris. As Professor Calvel's colleague Hubert Chiron has wryly pointed out, about the same time that his rural village in Brittany was discovering the "chic" Parisian baguette in the early 1960s, big city dwellers began clamoring for the charms of *pain de campagne*!

Figure 11–1 Country-Style Loaf Shapes.

Figure 11-2 One Long and One *Miche* Loaf.

Besides the traditional *boule*, *miche*, or *tourte*, and long loaves (Figure 11-2), quite a number of other country-style loaf shapes exist even today (see Figure 7-8 and Figure 7-9). Before the advent of the baguette and other *pains de fantaisie* caused loaves to become smaller and smaller (because they were generally sold whole and eaten the day they were baked), bread was made in round and oblong loaves weighing as much as 5 kilos (raw dough weight). These loaves kept for many days and were sold by weight. Smaller households bought pieces. Many of these are associated with particular regions, and were made with sourdough long before the introduction of bakers yeast.

---

rapid. The result is breads that are often overly dusted with flour, very large in volume, with a gray and somber-colored crumb, a pronounced *goût de bis* (taste of excessive flour ash) and a limited shelf life.

Furthermore, many country-style breads are made from ready-to-use, high-additive-content composite flour bases, with rapidity of production being the primary consideration. It should be pointed out that in order to better simulate the natural and rustic character of authentic *pain de campagne*, a considerable part of the industrially produced country-style breads offered to the French consumer has had its additive content further increased by the addition of calcium propionate.

Conversely, the formulas given in Exhibit 11-1 and Exhibit 11-2 permit the baker to ensure the rapid quality production of *pain de campagne* that respects and corresponds to the characteristics of the authentic high-quality *pain de campagne*.

Pain de Gruau

Until 1940, these products were made from *farine de gruau*,[2] a superior quality flour that had been made until the beginning of the 1920s with the fine grist obtained from the first break passages during the milling of common wheat. Since this was essentially a "skimming" of the best breadmaking flour, its production in this manner was forbidden by law. From that point the milling industry has been required to make *farine de gruau* by milling appropriately strong wheats.

As the production of traditional *farine de gruau* was forbidden from 1940 to 1956, the production of *pain de gruau* also ceased as a result and was difficult to revive when it was again permitted in 1956. The flours were of inappropriate quality, the procedures and techniques were badly understood, and these difficulties nearly caused it to be completely abandoned.

This type of production deserves to be better known, and it results in excellent quality breads. It is suitable for long baguettes, which are traditionally cut with 10 strokes of the blade, small hard rolls, three-cut "corkscrews," small "emperor" loaves (i.e., Kaiser rolls), "tobacco boxes," split loaves, and even "mushrooms."

The straight dough formula given in Exhibit 11-3 gives exceptional results, provided that a good-quality type 55 high protein *farine de gruau* is combined with systematic use of the autolysis rest period during mixing.[3] The dough is baked at a relatively low temperature, and baguettes and rolls should split evenly.

Rye Bread

Besides wheat, rye is the only cereal grain generally considered suitable for breadmaking (Figure 11-3). On a practical level, however, rye proteins do not have the same gluten-producing abilities as those of wheat. If the baker succeeds in making bread from pure rye flour, the production is often difficult and labor in-

---

Baguette Look-Alikes

An important note: *Pain de gruau* and *Pain viennois* look much like baguettes. It is important that bakers and their customers bear in mind that these loaves contain milk powder as well as butter or margarine and therefore are NOT baguettes. Consumer awareness of traditional lean-dough hearth breads has been too hard-won to confuse the issue. In the case of *Baguettes de Gruau*, for example, 10 slashes (as opposed to 5 to 7 for regular baguettes) are too subtle a differentiation for most customers. Careful labeling and knowledgeable sales staff must, in a positive way, make things clear.

Exhibit 11-1 Authentic Country Bread (*Pain de Campagne*) with Addition of Prefermented Dough

Formula

	Weight			Prefermented Dough	Dough
Wheat flour (type 55)	19.00 kg (670.19 oz)	95%		10.00 kg (352.7 oz)	9.00 kg (317.46 oz)
Rye flour (type 130)	1.00 kg (35.27 oz)	5%			1.00 kg (35.27 oz)
TOTAL FLOUR	20.00 kg (705.47 oz)	100%			10.00 kg (352.7 oz)
Water	13.80 kg (486.77 oz)	69%		6.9 kg (243.39 oz)	6.9 kg (243.37 oz)
Salt	400 g (14.11 oz)	2%		200 g (7.05 oz)	200 g (7.05 oz)
Yeast	400 g (14.11 oz)	2%		150 g (5.29 oz)	250 g (8.82 oz)
Malt extract	60 g (2.17 oz)	0.3%			60 g (2.17 oz)
Ascorbic acid	(400 mg)	(20 ppm)			(400 mg)
FERMENTED DOUGH				17.25 kg (608.47 oz)	17.25 kg (608.47 oz)
TOTAL DOUGH WEIGHT					34.66 kg (1222.57 oz)

Procedures

Mixing: Low speed	4 min	4 min
Mixing: (oblique axis) Second speed	6 min	8 min
Dough temperature from mixer	25° C / 77° F	25° C / 77° F
Fermented dough rest or storage time	5 h @ 24° C / 77° F *or* 16 h @ +4° C / 39° F	
First fermentation		40 min
Division, rest period		35 min
Dough molding		
Second fermentation / final proof		90 min
Baking @ 240° C / 464 ° F		30 min
PRODUCTION TIME		3 h 32 min

Exhibit 11-2 Authentic Country Bread (*Pain de Campagne*) by Sponge and Dough Method

Formula

	Weight			Yeast Sponge	Dough
Wheat flour (type 55)	19.00 kg (670.194 oz)	95%		6.00 kg (211.64 oz)	13.00 kg (458.55 oz)
Rye flour (type 130)	1.00 kg (35.27 oz)	5%			1.00 kg (35.27 oz)
TOTAL FLOUR	20.00 kg (705.47 oz)	100%		6.00 kg (211.64 oz)	14.00 kg (493.83 oz)
Water	13.80 kg (486.77 oz)	69%		3.54 kg (124.87 oz)	10.26 kg (361.90 oz)
Salt	400 g (14.11 oz)	2%			400 g (14.11 oz)
Yeast	200 g (7.05 oz)	1%		45 g (1.59 oz)	155 g (5.47 oz)
Malt extract	60 g (2.12 oz)	0.3%			60 g (2.12 oz)
Ascorbic acid	(400 mg)	(20 ppm)			(400 mg)
FERMENTED DOUGH				9585 g (338.09 oz)	9585 g (338.09 oz)
TOTAL DOUGH WEIGHT					34.46 kg (1215.52 oz)

Procedures

Mixing: Low speed	8 min	4 min
Mixing: (oblique axis) Second speed		8 min
Dough temperature from mixer	24° C / 75° F	24° C / 75° F
Sponge dough rest time	6 h	
First fermentation		100 min
Punch down after		(50 min)
Division, rest period		35 min
Dough molding		
Second fermentation / final proof		90 min
Baking @ 240° C / 464 ° F		35 min
PRODUCTION TIME		4 h 32 min

Exhibit 11–3 Straight Dough *Pain de Gruau* with Autolysis Rest Period

Formula

*Farine de gruau**	10 kg (352.73 oz)	100%
Water	6.9 kg (243.39 oz)	69%
Salt	200 g (7.05 oz)	2%
Yeast	200 g (7.05 oz)	2%
Lecithin or	15 g (0.53 oz)	0.15%
Margarine†	(25 g [0.88 oz])	(0.25%)
Milk powder†	150 g (5.29 oz)	1.5%
Malt extract	40 g (1.41 oz)	0.4%
Ascorbic acid	(100 mg)	(10 ppm)

Procedures

Mixing: 1st stage	4 min
Autolysis	17 min
Mixing: 2nd stage	9 min
Dough temperature	24° C / 75° F
1st fermentation	90 min
Punch down after	40 min
Divide, mold	35 min
2nd fermentation	75 min
Baking 240° C / 464° F	30 min
Total Dough Weight	17.53 kg
TOTAL TIME	4 h 20 min

*Note that French flour classifications, which are based on ash content, and American flour classifications, which are partially based on protein content, generally have very little in common. For that reason, it is difficult to make any statement regarding a simple equivalent between French and American flours. However, information on practical adjustments is included in the notes and text.

†Professor Calvel specifies (here and elsewhere) powdered milk rather than whole milk, and margarine rather than butter in the interest of making a stronger, more supple dough. In practice, good quality **unsalted** butter may be substituted, and one liter of whole milk may be substituted for each 100 g milk powder, along with a *readjustment of the quantity of water* called for in the formula. (Personal communication between Professor Calvel and James MacGuire, 1999)

Figure 11–3 Rye Loaves.

Rye loaves pose a particular challenge to bakers because the high pentosan content of rye flour prevents the formation of a gluten network. This is why rye bread recipes often call for a proportion of wheat flour.

In addition, the enzymes in rye flour remain active for a time after the starch gelatinizes during baking. These enzymes break down the starch in the bread with predictably mushy results. It is absolutely necessary to use a recipe involving a method of acidification or a sour to keep the enzymes in check.

Looser doughs yield best results, and kneading must be kept to the minimum necessary for gluten formation in order to avoid the breakdown of the small amount of gluten present. (See Exhibit 11–4 through Exhibit 11–7 for recipes).

tensive and the dough sticky and porous, causing the resulting product to be dense, heavy, and with low loaf volume.

For those reasons, the general practice used in France to facilitate production is to mix about 1/3 wheat flour with 2/3 rye flour. In certain other countries bakers mix wheat and rye flour on a half-and-half

Exhibit 11–4 Traditional Sourdough Rye Bread*

Refresher Culture (Rafraîchi)

Chef (starter)	300 g (10.58 oz)	
Medium rye flour	500 g (17.64 oz)	1,400 g (49.38 oz) of hydrated dough (This is a rather fluid dough)
Water	600 g (21.16 oz)	
Low speed mixing		Mix until smooth–there is no gluten formation
Dough temperature	25–26° C / 77–79° F	Depends on ambient conditions
Proof time	10 h	

*This is a new recipe for this North American edition, derived from the version given by Professor Calvel in the *Bulletin de l'Amicale des Anciens Élèves et Amis du Professeur Calvel*, No. 15, 1997.

Exhibit 11–5 Traditional Sourdough Rye Bread

Natural Fermented Culture (Levain)

Rafraîchi (Refresher Culture)	1400 g (49.38 oz)	
Medium rye flour	10.5 kg (370.37 oz)	17.4 kg (613.76 oz) of culture
Water	5.5 kg (194 oz)	
Low speed mixing		Mix until smooth dough–no gluten formation
Dough temperature	25–26° C / 77–79 ° F	
Proof time	7 h	

Exhibit 11–6 Traditional Sourdough Rye Bread

Dough Makeup (66% rye flour, 34% white flour)

Formula				Levain Addition	Total Ingredients
Bread flour*	6.8 kg (239.9 oz)				6.8 kg (239.86 oz)
Medium rye flour	2.2 kg (77.6 oz)			11 kg (388 oz)	13 kg (458.55 oz)
TOTAL FLOUR	8.8 kg (310.41 oz)	100%		11 kg (388 oz)	19.8 kg (670.19 oz)
(Total rye is 66%, wheat flour 34%)					
Water	7.3 l (257.5 oz)	69% overall		6.1 kg (215.17 oz)	13.86 kg (488.89 oz) (69%)
Chef (starter)					
Salt	360 g (12.7 oz)				360 g (12.7 oz) (1.8%)
Natural *levain* (sponge)				300 g (10.58 oz) (175 flour, 125 water)	
TOTAL DOUGH WEIGHT				17.4 kg (613.76 oz)	34.06 kg (1201.41 oz)

Procedures

Low speed mixing	12 min
Dough temperature	27° C/80° F
Division & rounding	10 min
Bench time / rest stage	20 min
Molding	10 min
Secondary fermentation	30 min
Baking (240° C / 464° F)	45 min

*Bakeries that frequently make considerable amounts of rye bread often use high-strength clear flours as the wheat portion of rye recipes because the darker color of clear flours is not a hindrance in rye breads.

---

basis, but in France this would result in what is termed *méteil*, or wheat-and-rye bread. The legal standard for the production and sale of rye bread in France requires a mix of 65% rye flour with 35% wheat flour.[4] When the amount of rye flour in the mix is less than 65%, the bread may no longer be legally called "rye bread," but must be identified as "bread with rye."

The quality of rye bread is improved when it is made from relatively acid doughs. For that reason, in those countries that are large-scale producers of rye bread, it is made from a natural *levain* without any addition of baker's yeast. However, when rye bread only represents a marginal part of total production, as is the case in France, it is more practical to use

Exhibit 11-7 Rye Bread with the Addition of Prefermented Dough and Baker's Yeast*

Formula

	Weight			Preferment Dough	Dough
Wheat flour† (type 55)	3.30 kg (116.4 oz)	33%		3.30 kg (116.4 oz)	
Gluten powder	200 g (7.05 oz)	2%			200 g (7.05 oz)
Rye flour (type 130)	6.50 kg (229.28 oz)	65%			6.50 kg (229.28 oz)
TOTAL FLOUR	10.00 kg (352.7 oz)	100%			6.70 kg (236.33 oz)
Water	6.90 kg (243.39 oz)	69%		2245 g (79.19 oz)	4655 g (164.2 oz)
Salt	200 g (7.05 oz)	2%		65 g (2.29 oz)	135 g (4.76 oz)
Yeast	200 g (7.05 oz)	2%		65 g (2.29 oz)	135 g (4.76 oz)
Wine vinegar‡	100 g (3.53 oz)	1%			100 g (3.53 oz)
Ascorbic acid	(200 mg)	(20 ppm)			(200 mg)
FERMENTED DOUGH				5675 g (200.18 oz)	5675 g (200.18 oz)
TOTAL DOUGH WEIGHT					17.4 kg (613.76 oz)

Procedures

Mixing: Low speed	4 min	5 min
Mixing: (oblique axis) Second speed	6 min	5 min
Dough temperature from mixer	25° C / 77° F	27° C / 80.6° F
Prefermented dough rest time	5 h @ 24° C / 77° F OR 18 h @ +4° C / 39° F	
First fermentation		50 min
Punch down after		(20 min)
Division, rest period		25 min
Dough molding		
Second fermentation / final proof		60 min
Baking @ 240° C		35 min
PRODUCTION TIME		3 h

*This is a good recipe for those bakers who do not make much rye bread, and who do not have much experience with the production of this somewhat involved and temperamental specialty product.

†Clear flours are often used in rye bread recipes for their strength, and because their darker colors are not an impediment.

‡It is possible to replace the red wine vinegar in this recipe with an appropriate amount of commercially available deactivated sourdough culture.

---

baker's yeast. Under those circumstances, it is wise to increase the acidity of the dough by incorporating the wheat flour portion of the formula in the form of a prefermented dough culture. In addition, during mixing the baker may add either 0.2% to 0.4% citric acid or 1% (on total flour weight) of vinegar to the dough.

The baker must also ensure that the rye flour used is not hyperdiastatic. A rye flour with about 1.2% ash content should be used in order to produce the best bread flavor. Finally, a small amount of wheat gluten powder (from 1.5% to 3.0% based on flour weight) may be added to the dough mixture in order to improve bread volume.[5]

It is important to avoid overmixing.[6] It is worth noting that the large proportion of *levain* accounts for the short primary and secondary fermentations. The oven temperature should be relatively high at the outset, and the loaves should be loaded into an oven containing a good deal of steam. The oven temperature should then be reduced, and the vent should be opened to remove the steam. Finish baking in a dry atmosphere. These loaves have excellent keeping qualities, and are best eaten after a 24-hour rest following baking.

Mixing should be stopped when the dough has formed a smooth mass. At that point it will have a slightly firm consistency, but not overly so. It is advisable to punch down the dough lightly during the first fermentation.

Either round loaves or short, only slightly elongated loaves may be formed after rounding or *la tourne*. The loaves are ready to bake when they have tripled in volume. Elongated loaves are improved by being moistened before being loaded into the oven, which should be injected with steam. They should also be sprayed or moistened with water when they are removed from the oven. Finally, rounded loaves are improved by being baked slightly longer than elongated loaves.

Benoîtons, which are small rye-raisin rolls, can be made by adding 160 g raisins to each kilogram of rye dough. These excellent small rolls are formed into a ball before proofing and baking.

Walnut Bread

Walnut breads are a relatively new product in much of France and only began to awaken the interest of French bakers in the early 1970s. They were known in Alsace before that time, but were made infrequently. Even today the production is relatively small, with most interest being shown by the restaurant trade, which serves walnut breads with cheeses. Walnuts are used with several different types of doughs, so the product range is rather varied.

In my opinion, the addition of walnuts seems to be most successful in the case of doughs made with *levain de pâte* or rye doughs. My preference is for mixed wheat and rye breads (Exhibit 11–8), which contain around 35% rye flour. This ratio produces bread which is remarkably compatible with all types of cheeses in terms of density, freshness, aroma, and keeping qualities.

Be sure that the walnuts used are in good, sound condition and are not rancid. Nuts are added to the dough near the end of mixing. The walnuts may all be chopped, or three quarters of them may be chopped, with the remaining quarter left intact. Whole walnuts are incorporated into the dough piece as it is being molded.

Sometimes the baker may wish to avoid mixing a special dough for walnut bread. In such cases it is possible to make a dough that is very close to that made from the formula above by mixing 1.1 kg of rye bread dough with 1 kg of baguette dough. The result is a dough which contains about 32% rye flour. Be

---

Exhibit 11–8 Mixed Wheat and Rye Bread with Walnuts

Formula

	Weight		Preferment Dough	Dough
Wheat flour (type 55)	3.40 kg (119.93 oz)	68%	1.50 kg (52.9 oz)	1.90 kg (67.02 oz)
Rye flour (type 130)	1.60 kg (56.44 oz)	32%		1.60 kg (56.44 oz)
Total flour	5.00 kg (176.37 oz)	100%		3.50 kg (123.46 oz)
Water	3.40 kg (119.93 oz)	68%	1020 g (35.98 oz)	2.38 kg (83.95 oz)
Salt	100 g (3.53 oz)	2%	30 g (1.06 oz)	70 g (2.47 oz)
Yeast	100 g (3.53 oz)	2%	30 g (1.06 oz)	70 g (2.47 oz)
Wine vinegar	50 g (1.76 oz)	1%		50 g (1.76 oz)
Ascorbic acid	(200 mg)	(20 ppm)		(200 mg)
Fermented dough			2580 g (91 oz)	2580 g (91 oz)
Walnuts	(150 g (5.29 oz) / kg of dough)			1300 g
	150 x 8.65 = 1297.5 = 1300 g			
TOTAL DOUGH WEIGHT				9.95 kg (350.97 oz)

Procedures

		Preferment Dough	Dough
Mixing: Low speed		4 min	3 min
Beater: Second speed		6 min	7 min
Dough temperature from mixer		25° C / 77° F	26° C / 78.8° F
Prefermented dough rest time		5 h @ 24° C / 77° F *or* 16 h @ +4° C / 39° F	
First fermentation			60 min
Division, rest period			15 min
Dough molding			15 min
Second fermentation / final proof			70 min
Baking @ 240° C			35 min
prick loaves: do not slash			
PRODUCTION TIME			3 h 25 min

Walnut, raisin, and raisin walnut breads are the most traditional flavored breads, and among the most interesting.

especially careful that the dough is homogeneous and the gluten network well formed. After mixing is completed, follow the above procedures from the first fermentation through baking.

Raisin Bread

These breads or rolls may be made with a rye-based dough, prepared as in the case of walnut bread. From 0.3% to 0.5% of dry malt or malt syrup may be added to the dough during mixing. Raisins that have been sorted and washed in advance may be added to the dough just at the end of mixing, according to the following ratios:

- raisin breads: 150 to 180 g per kilogram of dough
- walnut and raisin breads: 80 g chopped nuts per kilogram of dough; 80 g raisins per kilogram of dough.

Whole-Grain Bread and Bran Bread

These two products are actually very similar to one another when they are made up with the proper flours (Figure 11–4). In principle, whole-wheat bread (Exhibit 11–9) should result from the use of a whole-grain flour, including both the bran and the entire wheat berry. This is milled at an extraction rate of 98% to produce type 150 flour. Bran bread (Exhibit 11–10) is prepared from a mixture of 20% (minimum) straight wheat bran with 80% type 55 bread flour.

The bran used should not exceed the legally permitted tolerances for France, which are: 1 milligram per kilogram of flour for Lindane, and 2 mg per kilogram of flour for Malathion. It should be clearly understood by the consumer that in spite of all the care exercised in cleaning wheat before milling, consumption of both whole-wheat and bran breads, which have a high proportion of the outer bran layers of the wheat berry, may sometimes entail a risk of absorption of undesirable chemical residues. *These residues are absent from bread that is derived from standard white type 55 bread flour.*

As a practical consideration, it is common to add a little gluten powder to the flour during mixing in order to improve both dough tolerance and loaf volume. Furthermore, it should be noted that the development of the gluten matrix during mixing is difficult with such flours, and the baker should plan for a mixing time that is very slightly longer than average. Longer mixing will produce a smooth dough that "picks up" properly from the walls of the mixer bowl. A greater degree of fermentation activity than average will also be apparent. For that reason, the baker should expect

Figure 11–4 Whole-Wheat and Bran Loaves.

to use a slightly longer first fermentation than normal with these doughs, but a notably shorter second fermentation or proof.

Five-Grain Bread

Five-grain bread was traditionally a Swiss specialty until rediscovered in the 1970s by the French (and others). At about that time, the more progressive French millers began to become interested in it. Some of them—those with more inflationary tendencies—even advocate the use of six or seven grains.[7] This product is only made from a base or composite flour mix, and today it is common to find bakers who make this specialty product.

The mix of cereal flours used always includes wheat flour and rye flour, and a choice of flours made from barley, oats, buckwheat, soy, millet, and others. The wheat flour used is a slightly gray flour, a type 80 that is extracted from a relatively strong wheat or from a common wheat reinforced by a small amount of added gluten.

The flours of other cereals—or grasses—are extracted at a rate of 85% to 90%, using a rather coarse mill setting to obtain a slightly granular product somewhat like semolina, rather than very fine flours. This is because the fine flours made from cereals or grasses that are not normally used in breadmaking tend to further weaken the physical properties of doughs made from the flour mixes. Mixtures of several different flours might be considered, including those in Table 11–1.

General considerations on multigrain bread production include the following:

Exhibit 11-9 100% Whole-Wheat Bread

Formula

Prefermented Dough Method

Ingredient	Weight		Prefermented Dough	Dough
Whole-wheat flour	9.85 kg (347.44 oz)		1.50 kg (52.91 oz) (15%)	8.35 kg (294.53 oz)
Gluten (1.5%)	150 g (5.29 oz)		whole-wheat	150 g (5.29 oz)
TOTAL FLOUR	10.0 kg (352.73 oz)	100%		8.5 kg (299.8 oz)
Salt	200 g (7.05 oz)	2%	30 g (1.06 oz)	170 g (6 oz)
Yeast	200 g (7.05 oz)	2%	30 g (1.06 oz)	170 g (6 oz)
Ascorbic acid	(30 mg)	30 ppm		(300 ppm)
Water	7.4 kg (261.02 oz)	70%	1035 g (36.51 oz)	6365 g (224.5 oz)
Fermented dough			2595 g (91.53 oz)	2595 g (91.53 oz)
TOTAL DOUGH WEIGHT	17.8 kg (627.87 oz)			17.8 kg (627.87 oz)

Procedures

Mixing: Low speed		3 min	4 min	3 min
Second speed (oblique axis)		12 min	6 min	12 min
Dough temperature from mixer		24° C / 75° F	24° C / 75° F	24° C / 75° F
Prefermented dough rest			4 h @ 24° C / 77° F OR 16 h @ +4° C / 39° F	
First fermentation		70 min		40 min
Punch down after		(40 min)		
Division, rest period		20 min		25 min
Dough molding				
Second fermentation / final proof		60 min		65 min
Baking @ 235° C / 455° F		35 min		35 min
PRODUCTION TIME		3 h 20 min		3 h

Exhibit 11-10 20% Bran Bread

Formula

Prefermented Dough Method

Ingredient	Weight		Prefermented Dough	Whole-Wheat Dough
Wheat flour	7.9 kg (278.66 oz)		1.50 kg (52.91 oz) (15%)	6.4 kg (225.75 oz)
Gluten (1.5%)	100 g (3.53 oz)			100 g (3.53 oz)
Wheat bran	2.0 kg (70.55 oz)			2.0 kg (70.55 oz)
TOTAL FLOUR	10.0 kg (352.73 oz)	100%		8.5 kg (292.82 oz)
Salt	200 g (7.05 oz)	2%	30 g (1.06 oz)	170 g (6 oz)
Yeast	200 g (7.05 oz)	2%	30 g (1.06 oz)	170 g (6 oz)
Ascorbic acid	(300 mg)	30 ppm		(300 ppm)
Water	7.4 kg (261.02 oz)	72%	1035 g (36.51 oz)	6.24 kg (220.11 oz)
Fermented dough			2595 g (91.53 oz)	2595 g (91.53 oz)
TOTAL DOUGH WEIGHT	17.8 kg (627.87 oz)			17.8 kg (627.87 oz)

Procedures

Mixing: Low speed		3 min	3 min	3 min
Second speed (oblique axis)		12 min	6 min	12 min
Dough temperature from mixer		25° C / 77° F	24° C / 75°	25° C / 77° F
Prefermented dough rest time			4 h @ 24° C / 77° F OR 16 h @ +4° C / 39° F	
First fermentation		65 min		50 min
Punch down after		(30 min)		
Division, rest period		20 min		25 min
Dough molding				
Second fermentation / final proof		65 min		70 min
Baking @ 235° C / 455° F		35 min		35 min
PRODUCTION TIME		3 h 20		3 h 15 min

Table 11–1 Composite Flour Mixture Examples (%)

Flour Type	Mixture A	Mixture B	Mixture C
Wheat type 80	58.5	63	60
Powdered gluten	(1.5)	(2)	(2)
Rye flour	20	18	20
Barley flour	7	6	7
Oat flour	7	6	6
Millet flour	6	—	—
Buckwheat flour	—	5	—
Soy flour	—	—	5
TOTAL COMPOSITE FLOUR	100.0%	100.0%	100.0%

- Mixing, using an oblique-axis mixer, must not be excessive—3 minutes at first speed, followed by 7–8 minutes at second speed (the temperature of the dough may range up or down slightly around 24°C /75°F).
- The first fermentation, which lasts 60 minutes, will include two punching down stages at intervals of 20 minutes.
- The divided *pâtons*, which should weigh about 400 g (14.1 oz) each, are first rounded, then after a short rest are formed and deposited into previously greased cake pans or bread tins.
- The second fermentation will be relatively limited because of the limited tolerance of the dough, and will vary between 40 and 50 minutes.
- The dough pieces should be lightly misted with water before being placed in the oven.
- Baking is done at 230°C (446°F) for 35 minutes. The loaves are then depanned and placed on a baking sheet to bake for an additional 15 minutes.

The formula and procedures used may be either those for a traditional straight dough or a straight dough with the addition of prefermented dough, as shown in Exhibit 11–11 and Exhibit 11–12.

White Tinned Bread (United States and Canada— White Sandwich or Pullman Bread)

White bread —which has been made with a rich dough containing sugar, fat, and milk powder and then baked in a mold or bread pan—is the traditional bread of the English-speaking countries, and to a certain extent, of Japan as well.

This type of bread has traditionally been made by the sponge and dough method, called the *levain-levure* method in France. It is unfortunate that this tradition is often not practiced today, except in Japan, since its abandonment contributes to a lessening of bread quality. In any case, since the use of a prefermented yeast culture generally ensures the production of bread that really is superior quality in terms of taste, flavor, and keeping qualities, every effort should be made to encourage the retention and use of the somewhat similar (but less time-intensive) method based on the addition of prefermented dough (Exhibit 11–13). As in the case of traditional breads, makeup of the final dough is simplified by using a prefermented dough of the same type.

It should be noted in passing that the flour used in the production of tinned sandwich bread may be a type 55 flour with a slightly higher strength than the average (with a Chopin alveograph W reading around 220) and an ash content slightly less than normal, between 0.50% and 0.55%.

The formula in Exhibit 11–14 can also be used to make a very fine tinned sandwich bread or Pullman bread without the use of prefermented dough. However, in terms of keeping qualities, and even though milk does contribute to the overall quality, the taste of the bread will be inferior to that obtained than when it is made with prefermented dough.

The baker should note that pasteurized milk is preferable to long shelf life sterilized milk (common in France, but not in the United States or Canada) or canned milk in terms of taste. On the other hand, pasteurized milk must be more closely monitored, since acidified or soured milk damages the quality

Exhibit 11–11 Multigrain Bread: Traditional Straight Dough Method

Formula

Flour mix (B)	10 kg (352.73 oz)	100%
Salt	200 g (7.05 oz)	2%
Yeast	500 g (17.64 oz)	5%
Ascorbic acid	(300 mg)	(30 ppm)
Water	7 l (246.91 oz)	70%
TOTAL DOUGH WEIGHT	18.1 kg (638.45 oz)	

Procedures

Mixing: 1st speed (Oblique axis) 2nd speed	12 min
Dough temperature	24° C / 75° F
1st fermentation	60 min
Divide, form, turn	20 min
2nd fermentation	45 min
Baking 230° C / 446 ° F	35 + 15 min
PRODUCTION TIME	3 h 7 min

Exhibit 11-12 Multigrain Bread by the Prefermented Dough Method

Formula

	Weight		Prefermented Dough	Dough
Flour mix (B)	10.0 kg (352.73 oz)	100%	2.0 kg (70.55 oz) flour mix	8.0 kg (282.19 oz)
Salt	200 g (7.05 oz)	2%	40 g (1.41 oz)	160 g (5.64 oz)
Yeast	400 g (14.11 oz)	4%	40 g (1.41 oz)	360 g (12.7 oz)
Malt extract	40 g (1.41 oz)	0.4%		40 g (1.41 oz)
Ascorbic acid	(250 mg)	(25 ppm)		(250 mg)
Water	7.4 l (261.02 oz)	74%	1480 g (52.2 oz)	5.92 kg (208.8 oz)
PREFERMENT DOUGH			3.56 kg (125.57 oz)	3.56 kg (125.57 oz)
TOTAL DOUGH WEIGHT				18.04 kg (636.3 oz)

Procedures

	Prefermented Dough	Dough
Mixing: Low speed	10 min	12 min
(Oblique axis): Second speed		
Dough temperature from mixer	24° C / 75° F	24° C / 75° F
Prefermented dough rest time	3 h 30 min at 24° C / 75° F	
First fermentation		30 min
Division, rest, turn		20 min
Second fermentation		45–50 minutes
Baking @ 230° C		35 + 15 minutes
PRODUCTION TIME		2 h 42 min

Exhibit 11-13 Tinned Sandwich or Pullman Bread by the Prefermented Dough Method

Formula

	Weight		Prefermented Dough	Dough
Flour	10.0 kg (352.73 oz)	100%	5.0 kg (176.37 oz)	5.0 kg (176.37 oz)
Water	6.2 kg (218.67 oz)	62%	3.4 kg (119.93 oz)	2.8 kg (98.77 oz)
Salt	200 g (7.05 oz)	2%	100 g (3.53 oz)	100 g (3.53 oz)
Yeast	400 g (14.11 oz)	4%	100 g (3.53 oz)	300 g (10.58 oz)
Sugar	300 g (10.58 oz)	3%		300 g (10.58 oz)
Fat	450 g (15.87 oz)	4.5%		450 g (15.87 oz)
Milk powder	150 g (5.29 oz)	1.5%		150 g (5.29 oz)
Malt extract	50 g (1.76 oz)	0.5%		50 g (1.76 oz)
Ascorbic acid	(300 mg)	(30 ppm)		(300 mg)
PREFERMENT DOUGH			8.6 kg	8.6 kg (303.35 oz)
TOTAL DOUGH WEIGHT				17.75 kg (626.10 oz)

Procedures

	Prefermented Dough	Dough
Mixing: Low speed	3 min	4 min
(Beater): Second speed	6 min	8 min
Dough temperature from mixer	24° C / 75° F	27° C / 80.6° F
Prefermented dough rest time	5 h at 24° C / 75° F or 15 h at +5° C / 41° F	
First fermentation		25 min
Division, rest		25 min
Dough forming		
Second fermentation (as needed)		50–60 min
Baking @ 225° C		35–40 min
PRODUCTION TIME		2 h 42 min

Exhibit 11-14 Tinned Sandwich or Pullman Bread: Straight Dough Method with Added Milk

Formula

Wheat flour	10.0 kg (352.73 oz)	100%
Water	3.5 l (123.46 oz)	32%
Liquid milk	2.7 l (95.24 oz)	25%
Salt	200 g (7.05 oz)	2%
Yeast	400 g (14.11 oz)	4%
Malt extract	30 g (1.06 oz)	0.3%
Sugar	350 g (12.35 oz)	3.5%
Fat	450 g (15.87 oz)	4.5%
Ascorbic acid	(300 mg)	(30 ppm)
TOTAL DOUGH WEIGHT	17,430 g or 17.43 kg (614.81 oz)	

Procedures

Mixing: 1st speed	4 min
(Oblique axis): 2nd speed	8 min
Dough temperature	27° C / 80.6° F
1st fermentation	35 min
Divide & rest	25 min
Forming	
2nd fermentation	50–60 min
Baking	35–40 min
Production time	2 h 52 min

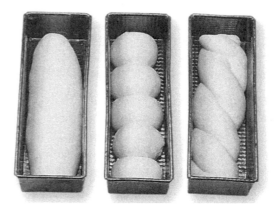

Figure 11-5 Unbaked Pullman Loaves.

of the dough in which it is used, as well as the end product.

Makeup Methods for Tinned Sandwich Breads

Dough used in the production of tinned sandwich breads may undergo several different makeup methods. These techniques ensure that the breads will have a smoothly homogeneous and fine-celled crumb structure.

Pâtons (unbaked loaves) may be made up as follows (Figure 11-5):

- The traditional cylindrical or "sausage" form, which is the most commonly used, is most effective when mechanical molding is used following a good, long rest period after dough division. Although care should be taken not to tear the dough, it should be given a thorough degassing through a dough brake or sheeter before it is rolled up, given final form, and placed in the baking pan;
- The "twist" form is accomplished by dividing the dough piece in halves, forming each half into a cylindrical *pâton* with half the diameter of the above method, and then twisting the two dough pieces together before placing them in the baking pan.
- The "multipart" method is accomplished by dividing the dough into several equally sized pieces, then carefully molding them into small cylindrical dough units. These are then placed side by side in a single row along the lengthwise axis of the baking pan.
- The similar "four-part" variation involves carefully making up a cylindrical or "sausage" shaped dough piece, which is then divided into sections from 6 to 7 cm long. As these dough pieces are placed in the baking pan they are twisted one quarter turn, so that the cut ends of the dough pieces face the long sides of the baking pan.[8]

Molding Pâtons: Relationship Between Dough Weight and Baking Pan Volume

The weight of the dough piece to be placed into a pan is a function of the volume of the baking pan and whether the pan is open-topped or lidded, as well as the density of the bread the baker wishes to produce. Whatever product or pan type is selected, it is important to always ensure that the *pâtons* are placed along the center line of the pan. (This will minimize the production of lopsided or misformed loaves.)

For medium-density breads, use the following "panned weight" guidelines:

- open pans: 260 g (9.173 oz) dough per liter of pan volume (0.5732 lbs of dough per each 61.025 cu in of pan volume, or 0.0939 lbs of dough for each 10 cu in of pan volume).
- lidded pans: 245 g (8.64 oz) dough per liter of pan volume (0.54 lbs of dough per each 61.025 cu in of pan volume or 0.088 lbs of dough for each 10 cu in of pan volume).

For higher-density breads, which are generally baked in lidded molds or forms, use approximately 275 g (9.7 oz) dough per liter of pan volume (0.606 lbs of dough per each 61.025 cu in of pan volume or 0.099 lbs of dough for each 10 cu in of pan volume).

Desired Appearance of Pan Breads after Baking

The baker has good reason to be concerned about this, since some types of pan bread have a tendency to become deformed through slight shrinkage or collapse of the sidewalls. It is important to thoroughly bake the product in order to avoid this problem (Figure 11–6). Furthermore, in the case of open-topped pans, the top crust should not rise excessively above the top edge of the baking pan. Overproofing increases the surface area that must be carried by the side walls of the loaf, while increasing their fragility and contributing to deformation and sidewall collapse or shrinkage.

Dough forming technique also plays a role: when the dough is panned as a simple cylinder, deformation in baking will be more pronounced than if the dough pieces have been "twisted" or made up according to one of the other methods described above.

The baker should avoid baking dough pieces that have risen too much. Overproofing results in loaves with excessively straight sidewall angles, a pronounced tendency toward loaf deformation,[9] and crowding of air cells within the crumb, along the interior wall of the bread crust.

Bread Pan Form and Characteristics

In baking tinned sandwich breads, it is useful to note that baking pans constructed of blued or blackened steel will become much darker in the course of the "tempering" bake that precedes their use in production baking. This dark coating facilitates heat conductivity, assists in the transmission of radiant heat, and markedly increases the anti-adhesive or "anti-stick" properties of the bread pans.[10]

As far as the proper shape for the bread pan is concerned, it is important to stress that the angles at the base of the pan should be slightly rounded on a radius of about 5 mm. This slight rounding, along with proper pan wall angles, greatly simplifies pan cleaning and maintenance, pan greasing, and removal of baked bread.

Brioche-Type Breads

As their name indicates, these breads are closely related to *brioches*. Although they are notably less rich than the latter, the dough (Exhibit 11–15 and Exhibit 11–16) from which they are made is required by French law to contain eggs at a level of at least 10% of the flour weight. It is the use of eggs at this proportion that justifies the term "brioche-type bread" in France. The fat used is generally an anhydrous fat, either a vegetable fat or a margarine, with a point of fusion ranging from 32 to 33°C / 89.6 to 91.4°F in winter and 35–36°C / 95 to 96.8°F in summer.

In place of using an artificial yellow egg color, as some professionals occasionally do, it is preferable to use a vegetable fat that is naturally colored by the inclusion of palm oil. This gives a lovely yellow color of natural origin to the bread crumb.

The flour used for this product should be a superior quality type 55 flour, with a W value equal or superior to 220 and a P/L ratio of 0.6 to 0.7. Leavening may be either by the straight dough method or by a pre-fermented culture, which as always will give consistently excellent results. This product is baked in open-topped pans, with the weight of the dough calculated on the basis of 230 to 240 g dough per liter of volume

---

The shaping method and whether or not covered molds are used have considerable effect upon the texture and crumb structure of the resulting loaves.

Figure 11–6 Baked Pullman Loaves.

Figure 11–7 Pain Brioché.

Pain brioché (Figure 11–7) is similar to *brioche* as its name indicates, but is less rich. It resembles challah or egg bread and in France is often served toasted as an accompaniment to *foie gras*.

Exhibit 11–15 Brioche-Type Bread: Straight Dough Method

Formula

High quality wheat flour	2.0 kg (70.55 oz)	100%
Water	960 g (33.86 oz)	45%
Salt	40 g (1.41 oz)	2.0
Yeast	80 g (2.82 oz)	4%
Sugar	300 g (10.58 oz)	15%
Milk powder	40 g (1.41 oz)	2%
Fat	800 g (28.22 oz)	8%
Eggs	200 g (7.05 oz)	10%
Ascorbic acid	(60 mg)	(30 ppm)

Procedures

Mixing: 1st speed	3 min
(Beater): 2nd speed	10 min
Dough temperature	26° C / 78.8° F
1st fermentation	45 min
Divide & Rest	20 min
Forming	
2nd fermentation	60 min
Baking 220° C / 428° F	30 min
Production time	2 h 48 min

Exhibit 11–16 Brioche-Type Bread by the Prefermented Dough Method (Pain Brioché)

Formula

	Weight			Prefermented Dough	Dough
Flour	2.0 kg (70.55 oz)		100%	800 g (28.22 oz) (40%)	1.2 kg (42.33 oz)
Water	960 g (33.86 oz)		45%	544 g (19.19 oz)	416 g (14.67 oz)
Salt	40 g (1.41 oz)		2%	16 g (0.56 oz)	24 g (0.84 oz)
Yeast	70 g (2.47 oz)		3.5%	16 g (0.56 oz)	54 g (1.90 oz)
Sugar	300 g (10.58 oz)		15%		300 g (10.58 oz)
Milk powder	40 g (1.41 oz)		2%		40 g (1.41 oz)
Malt extract	10 g (0.35 oz)		0.5%		10 g (0.35 oz)
Fat	160 g (5.64 oz)		8%		160 g (5.64 oz)
Eggs	200 g (7.05 oz)		10%		200 g (7.05 oz)
Ascorbic acid	(40 mg)		(20 ppm)		(40 mg)
PREFERMENT DOUGH				1376 g (48.54 oz)	1376 g (48.54 oz)
TOTAL DOUGH WEIGHT					3.78 kg (133.33 oz)

Procedures

Mixing: Low speed	3 min	4 min
(Beater): Second speed	10 min	8 min
Dough temperature from mixer	24° C / 75° F	26° C / 78.8° F
Prefermented dough rest time	4 h at 25° C / 77° F or 16 h at +4° C / 39 ° F	
First fermentation		30 min
Division, rest		20 min
Dough forming		
Second fermentation		70 min
Baking 220° C / 428° F		30 min
PRODUCTION TIME		2 h 42 min

Figure 11-8 Kaiser Rolls (petits pains Emperor)

Vienna loaves are often seen in *baguette* and *bâtard* shapes (not shown), but Kaiser rolls are the form best known in North America (Figure 11-8). These specialty rolls can be shaped either by hand or with the aid of a special tool.

of the bread pan (0.507 to 0.529 lbs. of dough per each 61.025 cu in of pan volume or 0.083 to 0.0867 lbs. of dough per each 10 cu in of pan volume).

Brioche-type bread dough is commonly used for making braided loaves and "pigtail" loaves, which are often baked in the Scandinavian countries.

Vienna Breads

In any discussion of the production of the Vienna-style baguettes and rolls produced in France today, it is important not to confuse them with the Vienna bread that was made there from 1840 to the 1920s and 1930s era. The Vienna bread made during that period was the result of a production method introduced into France by an Austrian, Baron Zang. With the assistance of a group of Viennese bakers, the Baron began to produce this type of bread in Paris in 1840, in a bakery that still exists today on the Rue de Richelieu.

This bread corresponded to our definition of "traditional" bread, since it was made from a mixture of flour, water, yeast and salt, occasionally enriched with the addition of a little malt extract. It was leavened with a *poolish*, and was thus a baker's yeast leavened product.

During the whole of the 1840s–1920s, this type of bread experienced well-deserved commercial success in the larger French cities, especially Paris. As may be seen still today on the advertising signs of old bakeries, the consumer had the choice between "French bread" made from a *levain*, and Vienna-style bread, made from baker's yeast. Throughout the 1920s, the use of the *poolish* method gradually began to decline in the face of increased use of yeast-leavened straight doughs, and authentic Vienna-style bread began to disappear.

The so-called Vienna breads and rolls that began to make an appearance toward the end of the 1940s partially replaced *farine de gruau*-based baguettes and rolls, since production of these latter products was essentially forbidden from August 1940 to September 1956.

Today's Vienna bread is far different from that of the 1840s–1920s. At present, the dough is enriched with fats, sugar, and milk powder. The dough pieces are molded and formed into cylindrical or "sausage" shape, the tops are slashed, and they are baked on baking screens or pierced baking trays. The crust is thin and often soft. Since leavening is often by the direct or straight dough method, the resulting product is often of barely acceptable quality.

However, when a portion of prefermented dough is added at mixing, as in the formula shown in Exhibit 11-17, it is possible to produce bread of excellent quality with superior aroma and keeping characteristics.

Milk Rolls

Milk rolls, as the name indicates, are made with a dough that has been moistened with milk (Exhibit 11-18). However, in France it is permissible to use water mixed with powdered nonfat milk. This moderately rich dough can be used to produce excellent products, provided that appropriate procedures are followed.[11]

This dough may also be used to make products in a variety of shapes: weavers' shuttles (small elongated rolls with pointed ends), small balls which are cut with scissors, and braids or pigtails. Dough for these products is placed on a baking sheet then washed with egg before baking.

The dough may be yeast leavened directly as a straight dough, or with the addition of a prefermented culture such as a sponge and dough yeast sponge or prefermented yeast dough. The mixed dough is of medium consistency.

Exhibit 11-17 "Vienna" Bread & Rolls by the Prefermented Dough Method

Formula

	Weight			Prefermented Dough	Dough
Flour	10.0 kg (352.73 oz)	100%		3.0 kg (105.82 oz) (30%)	7.0 kg (246.91 oz)
Water	6.8 kg (239.86 oz)	68%		2040 g (71.96 oz)	4.76 kg (167.9 oz)
Salt	200 g (7.05 oz)	2%		60 g (2.12 oz)	140 g (4.94 oz)
Yeast	300 g (10.58 oz)	3%		60 g (2.12 oz)	240 g (8.47 oz)
Malt extract	40 g (1.41 oz)	0.4%			40 g (1.41 oz)
Milk powder	200 g (7.05 oz)	2.0%			200 g (7.05 oz)
Sugar	250 g (8.82 oz)	2.5%			250 g (8.82 oz)
Fats	250 g (8.82 oz)	2.5%			250 g (8.82 oz)
Ascorbic acid	(300 mg)	(30 ppm)			(300 mg)
PREFERMENT DOUGH				5.16 kg (182 oz)	5.16 kg (182 oz)
TOTAL DOUGH WEIGHT					18.04 kg (636.3 oz)

Procedures

Mixing: Low speed		3 min	4 min
(Beater): Second speed		6 min	8 min
Dough temperature from mixer		24° C / 75° F	26° C / 78.8° F
Prefermented dough rest time		5 h at 24° C / 75° F or 15 h at +4° C / 39° F	
First fermentation			30 min
Division, rest			20 min
Dough forming			
Second fermentation (as needed)			45 min
Baking			25 min
PRODUCTION TIME			2 h 12 min

Exhibit 11-18 Milk Rolls

Formula

Milk Rolls by the Straight Dough Method			Milk Rolls by the Sponge & Dough Method

	Weight		Yeast Sponge	Dough
Flour	2.0 kg (70.55 oz)	100%	400 g (14.11 oz) (20%)	1.6 kg (56.44 oz)
Milk	1.2 kg (42.33 oz)	60%	260 g (9.17 oz)	940 g (33.16 oz)
Salt	40 g (1.41 oz)	2%		40 g (1.41 oz)
Yeast	60 g (2.17 oz)	3%	5 g (0.18 oz)	55 g (1.94 oz)
Butter	280 g (9.88 oz)	14%		280 g (9.88 oz)
Sugar	240 g (8.47 oz)	12%		240 g (8.47 oz)
Malt Extract	10 g (0.35 oz)	0.5%		10 g (0.35 oz)
Ascorbic acid	(40 mg)	(20 ppm)		(40 mg)
YEAST SPONGE			665 g (23.46 oz)	665 g (23.46 oz)
TOTAL DOUGH WEIGHT				3.83 kg (135.1 oz)

Procedures

Mixing: Low speed		3 min	5 min	3 min
(Beater): Second speed		6 min		6 min
Dough temperature from mixer		24° C / 75° F	25° C / 77° F	25° C / 77° F
Yeast sponge rest time			4 h	
First fermentation		80 min		45 min
Division, rest		15 min		15 min
Dough forming				
Second fermentation		65 min		65 min
Baking (215° C)		20 min	+	20 min
PRODUCTION TIME		3 h 9 min		2 h 34 min

Sandwich Rolls

These rolls are made from a dough (Exhibit 11–19) similar to that used to produce milk rolls. However, as they are intended to be eaten with sandwich fillings, including salted products, the formula is lower in sugar. In order to enhance the freshness and softness of the crumb, they are generally higher in butter or fat content as well. For the same reason, it is generally preferable to leaven them with a prefermented culture, or if using the straight dough method, to allow a longer first fermentation. It should also be noted that the weight of the dough pieces (from 30 to 35 g) is lower than that used for milk rolls (45 g).

"Pistol" Rolls

This product is sometimes found in the northern and eastern parts of France, and especially in Belgium. The formula (Exhibit 11–20) is related to that of milk rolls, but is less rich. Provided that makeup procedures are appropriate, the quality of "pistol" rolls is excellent and often superior to that of milk rolls.

Leavening may be either by the straight dough method or by addition of a prefermented yeast sponge. The latter has the advantage of improving both the taste and keeping qualities and also ensures that the quality of product will be exceptionally regular from one batch to another.

Makeup of pistol rolls is carried out in two stages: the dough pieces are weighed out at 45 g each, rounded, covered to prevent surface drying, and allowed to rest for about 25 minutes with the molding seam on the bottom. At the end of this first interval, an oiled wand (comparable in diameter to a pencil or a chopstick) is pressed across the middle to *almost* divide the dough ball into two equal parts.[12] After this partial splitting, the dough balls are inverted onto a linen canvas with the split again on the bottom position. After a second proof period of about 30 minutes they are again flipped over so that the split is on top before being loaded into a steam-injected oven and baked at a fairly low temperature.

Gluten Bread

Current law in France requires gluten breads to be protein enriched: when sold as fresh bread, the protein enrichment level must be at least 14% by weight; when in the form of rusks, protein level must be at

---

Exhibit 11–19 Sandwich Rolls

Formula

	Sandwich Rolls by the Straight Dough Method			Sandwich Rolls by the Sponge & Dough Method	
	Weight			Yeast Sponge	Dough
Flour	2.0 kg (70.55 oz)	100%		400 g (14.11 oz) (20%)	1.6 kg (56.44 oz)
Milk	1.2 kg (42.33 oz)	60%		260 g (9.17 oz)	940 g (33.16 oz)
Salt	40 g (1.41 oz)	2%			40 g (1.41 oz)
Yeast	50 g (1.76 oz)	2.5%		5 g (0.18 oz)	45 g (1.59 oz)
Butter	320 g (11.29 oz)	16%			320 g (11.29 oz)
Sugar	60 g (2.12 oz)	3%			60 g (2.12 oz)
Malt extract	10 g (0.35 oz)	0.5%			10 g (0.35 oz)
Ascorbic acid	(40 mg)	(20 ppm)			(40 mg)
YEAST SPONGE				665 g (23.46 oz)	665 g (23.46 oz)
TOTAL DOUGH WEIGHT					3.680 kg (129.81 oz)

Procedures

Mixing: Low speed	3 min		5 min	3 min
(Beater): Second speed	6 min			6 min
Dough temperature from mixer	24° C / 75° F		25° C / 77° F	25° C / 77° F
Yeast sponge rest time			4 h	
First fermentation	120 min			60 min
Division, rest	15 min			15 min
Dough forming				
Second fermentation	65 min			70 min
Baking (230° C)	20 min			20 min
PRODUCTION TIME	3 h 50 min			2 h 55 min

Exhibit 11-20 Pistol Rolls

Formula

"Pistol" Rolls by the Sponge and Dough Method

	Weight		Yeast Sponge	Dough
Flour	2.0 kg (70.55 oz)	100%	500 g (17.64 oz) (20%)	1.5 kg (52.9 oz)
Milk	1.08 kg (38.09 oz)	54%	300 g (10.58 oz)	900 g (31.75 oz)
Salt	40 g (1.41 oz)	2%		40 g (1.41 oz)
Yeast	50 g (1.76 oz)	2.5%	10 g (0.35 oz)	40 g (1.41 oz)
Milk powder	40 g (1.41 oz)	2%		40 g (1.41 oz)
Margarine	200 g (7.05 oz)	10%		200 g (7.05 oz)
Sugar	70 g (2.47 oz)	3.5%		70 g (2.47 oz)
Malt extract	10 g (0.35 oz)	0.5%		10 g (0.35 oz)
Ascorbic acid	(40 mg)	(20 ppm)		(40 mg)
YEAST SPONGE			810 g (28.57 oz)	810 g (28.57 oz)
TOTAL DOUGH WEIGHT				3.61 kg (127.34 oz)

(A) sandwich rolls, (B) pistol rolls, (C) sugar-sprinkled milk rolls

Procedures

Mixing: Low speed	3 min	5 min	3 min
(Beater): Second speed	6 min		6 min
Dough temperature from mixer	24° C / 75° F	25° C / 77° F	25° C / 77° F
Yeast sponge rest time		4 h	
First fermentation		90 min	50 min
Divide, round, rest		35 min	35 min
Dough forming		5 min	5 min
Second fermentation		30 min	30 min
Baking (230° C)		20 min	20 min
PRODUCTION TIME		3 h 10 min	2 h 29 min

---

least 20% by weight. The formula given in Exhibit 11-21 will permit the baker to both meet these legal requirements and to ensure the production of a quality product.

Gluten breads have a high percentage of both water and protein and are recommended for diabetics. These products are also lower in calories than traditional breads.

Weigh out the *pâtons* during dough division so that the loaves, which are generally baked in a tin, do not rise more than 2 cm above the rim of the tin or form. This will help to avoid deformation of the loaves after baking. The dough pieces may be molded either into sausage shapes or into balls during forming. Bake the loaves thoroughly, but at the same time avoid excessive crust coloration.

BREADS FOR FILLING AND TOPPING

Various types of pizza doughs are well known. The *pissaladière*, as it is made in southern Provence and in the region of Nice, is not as well known but is certainly worth a brief description.

Exhibit 11-21 High-Quality Gluten Bread

Formula

Wheat flour type 55	4.25 kg (149.9 oz)	
Vital wheat gluten	750 g (26.46 oz)	
Total flour	5.00 kg (176.37 oz)	100%
Water	3.65 kg (128.75 oz)	70%
Salt	75 g (2.65 oz)	1.5%
Yeast	250 g (8.82 oz)	3%
Margarine	25 g (0.88 oz)	0.5%
Milk powder	25 g (0.88 oz)	0.5%
Ascorbic acid	(20 mg)	(5 ppm)
TOTAL DOUGH WEIGHT	9.025 kg (18.34 oz)	

Procedures

Mixing: 1st speed	5 min
Beater: 2nd speed	10 min
Dough temperature	25° C / 77° F
1st fermentation	30 min
Divide, rest	20 min
Forming	
2nd fermentation	50 min
Baking 230° C / 446° F	30 min
TOTAL TIME	2 h 25 min

The dough (Exhibit 11–22) is first rolled out onto a baking sheet. It is then covered with a 3 mm thick layer of finely chopped, sautéed onions lightly seasoned with salt and pepper. Anchovy fillets that have been briefly washed, rinsed, and dried are then applied in a diagonal cross pattern (Figure 11–9), and thin sticks of dough are arranged horizontally and vertically on the dough sheet to form squares, so that each square contains two anchovy fillets in an "x" pattern and two pitted black olives. After applying the toppings, allow the dough to rise for about 25 minutes, then sprinkle the *pissaladière* with olive oil before baking at around 225°C. After baking and cutting, serve it warm.

The base dough for *pissaladière* may also be used to make *fougasse* or fougasa bread, which is very well known in the southeast of France. This product is made from a rolled-out round or oval dough pancake from which small boat-shaped designs have been cut with a dough cutter. The "flambade" or "tongue of flame" bread, which was commonly made in the southwest part of France, is made from common bread dough but is baked in an extremely hot oven.

SAVORY AND AROMATIC BREADS

In the past, savory or aromatic breads were encountered from time to time, but they were strictly regional in character. For the most part, they were unknown to the majority of bakers, or were well known in one location while being of only marginal importance in another. During the early 1980s, however, the popularity of these breads began to spread, and while the total production remains rather small, the number of bakers who are interested in them continues to grow.

The Production of Savory Breads

As in the following cases, a number of breads based on aromatic or savory products may be produced from a single dough recipe (Figure 11–10). This base is a straight dough to which a relatively large proportion of prefermented dough is added in order to produce quality breads with good shelf life (Exhibit 11–23). This prefermented dough addition is a traditional bread dough that has not been overmixed and that has had a minimum 3-hour fermentation period.

Exhibit 11–22 Doughs for Filling and Topping (for example, Pizza, *Pissaladière*)

Formula

Bread Doughs for Topping by the Prefermented Dough Method

	Weight		Prefermented Dough	Dough
Flour	2.0 kg (70.55 oz)	100%	400 g (14.11 oz)	1.6 kg (56.44 oz)
Salt	36 g (1.27 oz)	1.8%	8 g (0.28 oz)	28 g (0.98 oz)
Yeast	40 g (1.41 oz)	2%	8 g (0.28 oz)	32 g (1.13 oz)
Malt extract	8 g (0.28 oz)	0.4%		8 g (0.28 oz)
Olive oil	60 g (2.17 oz)	3%		60 g (2.17 oz)
Ascorbic acid	(20 mg)	(10 ppm)		(20 mg)
Water	1.2 l (42.33 oz) (kg)	60%	272 g (9.59 oz)	928 g (32.73 oz)
PREFERMENT DOUGH			688 g (24.27 oz)	688 g (24.27 oz)
TOTAL DOUGH WEIGHT				3344 g or 3.344 kg (7.37 lbs)

Procedures

Mixing: Low speed	3 min	3 min
(Beater): Second speed	6 min	6 min
Dough temperature from mixer	24° C / 75° F	25° C / 77° F
Prefermented dough rest time	4 h at 24° C / 75° F or 18 h at +5° C / 41° F	
First fermentation		30 min
Division, dough forming		20 min
Second fermentation (as needed)		25 min
Baking @ 225° C		30 min
PRODUCTION TIME		1 h 55 min

Figure 11-9 Pizza and *Pissaladière*.

Professor Calvel's recipe calls for a firm dough for pizza and *pissaladière* because many bakeries in France bake these products on sheet pans and sell them by the portion. Individual and smaller round pizzas can be made with a more hydrated dough, as seen in Italy. The *Associazione Vesace Pizza Napoletona* would take issue with the addition of salt and olive oil, even though this would seem to be of little consequence. The addition of *pâte fermentée* is also not traditional, but the editors feel that it is an improvement.

Figure 11-10 Savory Loaves (Onion and Olive).

It may even be from a previous day's production, kept under refrigeration. The addition of the prefermented dough to the straight dough speeds up the breadmaking process while ensuring the production of high-quality breads.

Characteristics of Raw Materials

In all cases the flour should be a good quality wheat flour without additives[13] and with a W value between 180 and 200. The protein content should be at least 11%, and the flour should have a balanced enzymatic profile. The addition of a small amount of rye flour—about 5%—will tend to improve both the taste and the keeping qualities of the bread. Although these quality and strength characteristics are above average, they are not excessive.[14]

Addition of malt extract is not always necessary. However, whenever a fermented culture such as prefermented dough or a *poolish* is incorporated into the dough during mixing, it is often advisable to add malt extract to remedy the low sugar content of the fermented culture. However, an excess of malt extract tends to damage the physical properties of the dough and to produce an excessively reddish crust.

Furthermore, there is good reason to assist the maturation of the dough so that it will be able to better carry the weight of the aromatic ingredients. This is especially true for those bread types in which the dough pieces are baked in a mold or tin.

Finally, it should be noted that the addition of aromatic ingredients should always take place at the end of mixing, and that in the case of certain ingredients, the bowl, beater, and mixer should be thoroughly washed after use.[15]

Anise Bread

Anise is a relatively strong flavoring agent. It should be added at a rate of 60 g (2.12 oz) per kilogram of flour, or 120 g (4.23 oz) to be added to the basic dough batch given in the chart above. Anise is especially suitable for the production of small variety rolls.

Olive Bread

Plan to use 120 g (4.23 oz) pitted, finely chopped olives per kilogram of dough, taking care not to crush the olives. The bread is made up as small rolls or as *joko* loaves of 250 g (8.82 oz) each. They are appropriate to accompany salads and raw vegetables.

Exhibit 11–23 Base Formula for Aromatic Breads

Formula

Straight Dough With Addition of Prefermented Dough

Ingredient	Weight		Preferment Dough	Dough
Wheat flour	1.9 kg (67 oz)		360 g (12.7 oz)	1.54 kg (54.32 oz)
Rye flour	100 g (3.53 oz)		—	100 g (3.53 oz)
Total flour	2.0 kg (70.55 oz)	100%		1.64 kg (57.85 oz)
Salt	40 g (1.41 oz)	2%	7 g (0.25 oz)	33 g (1.16 oz)
Yeast	45 g (1.59 oz)	2.25%	8 g (0.28 oz)	37 g (1.31 oz)
Malt extract	10 g (0.35 oz)	0.5%		10 g (0.35 oz)
Ascorbic acid	(60 mg)	(30 ppm)		(60 mg)
Water	1400 g (49.38 oz)	70%	250 g (8.82 oz)	1.15 kg (40.56 oz)
FERMENTED DOUGH			625 g (22.05 oz)	625 g (22.05 oz)
TOTAL DOUGH WEIGHT				3.495 kg (123.28 oz)

Procedures

Mixing: Low speed	4 min	3 min
Second speed (beater)	3 min	5 min
Dough temperature from mixer	25° C / 77° F	25° C / 77° F
Prefermented dough rest time	3 h @ 25° C / 77° F OR 16 h @ +4° C / 39° F	
First fermentation		50 min
Division, rest period		30 min
Dough molding		
Second fermentation / final proof		75 min
Baking @ 235° C / 455° F		20 to 35 min
PRODUCTION TIME		3 h 20 min

Cumin Bread

Cumin seed is mixed into the dough at the rate of 20 g (0.71 oz) per kilogram of flour. This bread is especially suitable for consumption with strong cheeses.

Thyme Bread

Use 10 g (0.35 oz) thyme, chopped very small, per kilogram of flour. This bread is said to be proper for serving with wild game.

Garlic Bread

Add 80 g (2.82 oz) peeled and chopped garlic for each kilogram of flour. This bread is a suitable accompaniment for raw vegetables or red meats. It may be eaten grilled or toasted, with salads and thick soups.

Onion Bread

Prepare 500 g (17.64 oz) peeled and finely chopped onions for each kilogram of flour. Brown them over low heat, then let them drain and cool. Coat them with flour before adding them to the dough. The baker may also use 100 g (3.53 oz) of dehydrated onions, which are sautéed over low heat before use.

Often preceded by an unexpected tear in the eye, onion bread simply must be tasted. It is universally esteemed.

Fennel Bread

Fifteen grams of fennel should be added per kilogram of flour; fennel bread is especially popular with fish dishes.

Poppyseed Bread

Use 140 g (4.94 oz) lightly grilled poppyseed per kilogram of flour. The poppyseed is sprinkled on the moistened dough pieces before baking. The resulting bread is a perfect accompaniment to mixed salads and raw vegetables and may also be considered as a remedy for insomnia.

Cheese Breads

Auvergne blue cheese bread: Allow 150 g (5.3 oz) Auvergne blue cheese, coarsely chopped, for each kilogram of flour. Mix it into the dough just at the end of mixing. Although it must be evenly distributed throughout the dough, be careful not to reduce it to crumbs in the process. The resulting bread is eminently suitable for serving with mixed salads and raw vegetables.

Parmesan bread: Divide a 140 g (4.94 oz) block of Parmesan into two portions, one of about 80 g (2.82 oz), and the other of 60 g (2.12 oz). Grate the 80 g portion before adding it to the dough just at the end of dough mixing. Divide the 60 g portion into small cubes and insert them into the center of small round dough pieces. These small rolls are very popular with meats and good wines.

Algae or Seaweed Bread

Allow 100 g (3.53 oz) powdered algae (or seaweed) per kilogram of flour. Two different types of powdered seaweed are generally to be found: one is green and often very finely powdered; the other type, which is less widely known, is dark brown and less finely powdered. I prefer the latter.

The production of seaweed bread requires special care in mixing, as well as the inclusion of the prefermented dough produced according to the base formula given previously. The powdered seaweed should always be added at the *beginning* of mixing because it—especially when it is very finely ground—produces a notable softening of the dough, and the dough will be less hydrated.

The dietary value of seaweed bread may be significant because of its high mineral salts content. In addition, it might be considered especially appropriate for those who practice the underwater sport of spear fishing.

Sunflower Oil Bread with Sesame Seeds

These breads (Exhibit 11–24) are made up as small round rolls or *joko* loaves, weighing 350 g (12.35 oz) in dough form. After forming, they may be placed on a baking sheet or on baking screens. They are then razor-cut "en saucission," like Vienna baguettes. Before being placed in the oven, they are lightly sprayed with water and sprinkled with sesame seeds. These breads are high in fat, and although they are quite good-tasting, they should be eaten only in moderation.

Wheat Germ Bread

Add 150 g (5.29 oz) of wheat germ to each kg of flour at the beginning of mixing (Exhibit 11–25). Take care

Exhibit 11-24 Sunflower Oil Bread with Sesame Seeds

Formula

Wheat flour	1400 g (49.38 oz)	
Farina flour	600 g (21.16 oz)	
TOTAL FLOUR	2000 g (70.55 oz)	100%
Salt	40 g (1.41 oz)	2%
Yeast	50 g (1.76 oz)	2.5%
Malt extract	12 g (0.42 oz)	0.6%
Sunflower oil	240 g (8.47 oz)	12%
Water	1100 g (38.8 oz)	48%
Ascorbic acid	(20 mg)	(20 ppm)
PREFERMENTED DOUGH*	250 g (8.82 oz)	12.5%
TOTAL DOUGH WEIGHT	Dough weight 3692 g + prefermented dough = 250 grams = 3942 g (139.05 oz)	

Procedures

Mixing: 1st speed	3 min
Beater: 2nd speed	6 min
Dough temperature	25° C / 77° F
1st fermentation	50 min
Divide, rest	16 min
Forming	
2nd fermentation	75 min
Baking 240° C / 464°F	35 min
PRODUCTION TIME	3 h 5 min

*Note that the prefermented dough is not included in the flour or water of the recipe weight of the French edition.

Exhibit 11-25 Wheat Germ Dietetic Bread

Formula

Wheat flour	1.9 kg (67 oz)	
Medium rye flour	100 g (3.53 oz)	
Total flour	2.0 kg (70.55 oz)	100%
Wheat germ	300 g (10.58 oz)	15%
Salt	45 g (1.59 oz)	2.25%
Yeast	50 g (1.76 oz)	2.5%
Malt extract	10 g (0.35 oz)	0.5%
Ascorbic acid	(40 mg)	(20 ppm)
Water	1.32 kg (46.56 oz)	60%
PREFERMENTED DOUGH*	500 g (17.64 oz)	25%
TOTAL DOUGH WEIGHT	4225 g = 4.225 kg (149.02 oz)	

Procedures

Mixing: 1st speed	3 min
Beater: 2nd speed	6 min
Dough temperature	25° C / 77° F
1st fermentation	45 min
Divide, rest	16 min
Forming	
2nd fermentation	70 min
Baking 220° C / 428 ° F	30–40 min
PRODUCTION TIME	3 h

*Note that the prefermented dough is not included in the flour or water of the recipe weight of the French edition.

to use wheat germ that is in good, properly stored condition. When wheat germ has been improperly stored, there is a risk that it will add a detrimental degree of acidity to the dough, with an accompanying rancid taste. The prefermented dough portion of the base dough should be proofed for between 4 and 5 hours.

Wheat germ bread is a valuable dietetic food because of its high level of vitamin E, as well as its valuable mineral content. In addition, it has a very pleasant taste.

Bacon Bread

Use 125 g (4.41 oz) (per kilogram of flour) smoked bacon with the rind removed. Dice the bacon into small cubes, then sauté it briefly in a frying pan. Put it aside to cool at room temperature. The diced bacon will be added to the dough at the end of the mixing cycle, while the cooled bacon drippings will be added to the dough at the beginning of mixing (Exhibit 11–26).

Bread made in this fashion is good to serve with mixed salads and is especially suitable for slicing and grilling or toasting.

Zucchini (Vegetable Marrow) Bread

Allow 800 g (28.22 oz) zucchini squash (with stems removed) for each kilogram of flour. Grate the squash into matchstick-size pieces. Place in a colander in several layers, sprinkling each layer with part of the formula salt. After allowing the salt to draw moisture from the grated zucchini for 40 minutes, press down the layers in the colander to express the juice.

Mix together the flour, grated zucchini, the remainder of the salt, the yeast, and part of the formula water, which should be slightly warmed in advance (Exhibit 11–27). The amount of water added should be sufficient to make up a *pâte bâtarde* (a medium textured dough). Protect the dough from drafts and from the formation of a "skin" throughout the entire makeup procedure.

Form the dough into *joko* loaves. Just before placing them into the oven, brush them with a little olive oil to prevent the exposed pieces of zucchini from burning. After baking, allow the loaves to cool adequately on a grate or wire rack.

Pumpkin Bread

For each kilogram of flour, cook 850 g (29.98 oz) pumpkin pulp over low heat in water with a small

---

Exhibit 11–26 Smoked Bacon Bread

Formula

Wheat flour	1.9 kg (67 oz)	
Rye flour	100 g (3.53 oz)	
Total flour	2.0 kg (70.54 oz)	100%
Salt	36 g (1.27 oz)	1.8%
Yeast	50 g (1.76 oz)	2.5%
Malt extract	10 g (0.35 oz)	0.5%
Ascorbic acid	(40 mg)	(20 ppm)
Water	1.36 kg (47.97 oz)	68%
Diced smoked bacon	250 g (8.82 oz)	12.5%
Prefermented dough*	400 g (14.11 oz)	20%
TOTAL DOUGH WEIGHT	4106 g or 4.106 kg (144.83 oz)	

Procedures

Mixing: 1st speed	3 min
Beater: 2nd speed	6 min
Dough temperature	25° C / 77° F
1st fermentation	45 min
Divide, rest	16 min
Forming	
2nd fermentation	75 min
Baking 220° C / 428° F	30–40 min
PRODUCTION TIME	3 h 5 min

*Note that the prefermented dough is not included in the flour or water of the recipe weight of the French edition.

Exhibit 11–27 Zucchini (Vegetable Marrow) Bread

Formula

Wheat farina flour	2.0 kg (70.55 oz)	100%
Yeast	60 g (2.12 oz)	3%
Salt	60 g (2.12 oz)	3%
Grated zucchini	1600 g (56.44 oz)	80%
Olive oil	40 g (1.41 oz)	2%
Water	900 g (31.75 oz)	45%
TOTAL DOUGH WEIGHT	4.66 kg (164.37 oz)	

Procedures

Mixing: 1st speed	3 min
Beater: 2nd speed	6 min
Dough temperature	25° C / 77° F
1st fermentation	90 min
Punch down after	40 min
Divide / form	15 min
2nd fermentation	45 min
Baking (225° C)	35 min
PRODUCTION TIME	4 h 14 min

Exhibit 11-28 Pumpkin Bread

Formula

Wheat farina flour	2.0 kg (70.55 oz)	100%
Yeast	60 g (2.12 oz)	3%
Salt	50 g (1.76 oz)	2.5%
Pumpkin pulp	1700 g (59.96 oz)	85%
Cooking water	120 g (4.23 oz)	6%
TOTAL DOUGH WEIGHT	3.93 kg	

Procedures

Mixing: 1st speed	3 min
Beater: 2nd speed	6 min
Dough temperature	25° C / 77° F
1st fermentation	100 min
Punch down after	60 min
Divide / form	15 min
2nd fermentation	50 min
Baking (225° C)	35–40 min
PRODUCTION TIME	3 h 35 min

---

amount of salt for 25 to 30 minutes. Strain the pulp to homogenize it well. Allow it to cool, reserving the cooking water.

Dissolve the yeast in a small amount of the cooled cooking water. After a few minutes, add the cooled pumpkin pulp, the flour, and the salt, and begin mixing (Exhibit 11-28).[16] If more water is required to make up a *bâtarde* dough, sprinkle in a small amount of the cooking water. Divide and form the dough into balls and incise them in a circular pattern just before baking. After baking, allow the loaves to cool completely on a grate or screen.

Apple Bread

For each kilogram of flour, allow 750 g (26.46 oz) firm fleshed apples, preferably with a slightly bitter taste, such as the Granny Smith variety. Remove the seeds, peel the apples, and cut them into cubes. Sauté them lightly with a little butter over low heat, until they are slightly golden. Then mash them coarsely with a fork and allow them to cool.

Then proceed to mix the dough (Exhibit 11-29), which should be very lightly flavored with cinnamon. The strength of the dough will be improved by the long second fermentation or proof.[17]

Sausage Bread

Use 250 g all-pork Toulouse sausage[18] for each kilogram of flour. Cut the sausage into slices 1 cm thick. Sauté them for 10 minutes over low heat with a little olive oil, until they are slightly golden, then set side and allow to cool.

Mix a slightly firm dough (Exhibit 11-30), using the rest of the ingredients and the prefermented dough.

Exhibit 11-29 Apple Bread by the Poolish Sponge Method

Formula

	Weight		Poolish Sponge	Dough
Farina flour	2.00 kg (70.55 oz)	100%	400 g (14.11 oz)	1600 g (56.44 oz)
Salt	40 g (1.41 oz)	2%		40 g (1.41 oz)
Yeast	60 g (2.12 oz)	3%	20 g (0.7 oz)	40 g (1.41 oz)
Mashed apples	1500 g (52.91 oz)	75%		1500 g (52.9 oz)
Cinnamon	(hint)		—	(hint)
Water	(as needed)		400 g (14.11 oz)	
POLISH SPONGE WEIGHT			820 g (28.9 oz)	820 g (28.9 oz)
TOTAL DOUGH WEIGHT				3.6 kg (126.98 oz)

Procedures

Mixing: Low speed	3 min	3 min
Beater: Second speed		6 min
Dough temperature from mixer	25° C / 77° F	25° C / 77° F
First fermentation (ambient temperature)	(3–4 h)	90 min
Punch down after		(50 min)
Division, dough piece molding		16 min
Second fermentation / final proof		60 min
Baking @ 225° C		35–40 min
PRODUCTION TIME		3 h 35 min

Exhibit 11-30 Toulouse Sausage Bread by the Prefermented Dough Method

Formula

	Weight		Prefermented Dough	Final Dough
Wheat flour	1900 g (67 oz)		260 g (9.17 oz) (13%)	1640 g (57.85 oz)
Rye flour	100 g (3.53 oz)			100 g (3.53 oz)
Total flour	2000 g (70.55 oz)	100%	260 g (9.17 oz)	1740 g (61.38 oz)
Yeast	80 g (2.82 oz)	4%	6 g (0.21 oz)	74 g (2.61 oz)
Salt	30 g (1.06 oz)	1.5%	6 g (0.21 oz)	24 g (0.85 oz)
Sausage	500 g (17.64 oz)	25%		500 g (17.64 oz)
Olive oil	60 g (2.12 oz)	3%		60 g (2.12 oz)
Water	1240 g (43.74 oz)	55%	180 g (6.35 oz)	1060 g (37.39 oz)
Malt extract	14 g (0.49 oz)	0.7%		14 g (0.49 oz)
PREFERMENT DOUGH			452 g	452 g (15.94 oz)
TOTAL DOUGH WEIGHT				3924 g (138.41 oz)

Procedures

Mixing: Low speed	3 min	3 min
Beater: Second speed	10 min	6 min
Dough temperature from mixer	25° C / 77° F	25° C / 77° F
Prefermented dough rest time	4 h at 24° C / 75° F or 16 h at +4° C / 39° F	
First fermentation		40 min
Division, rest period		15 min
Dough forming		
Second fermentation		60 min
Baking @ 230° C		40 min
PRODUCTION TIME		2 h 45 min

When mixing has been completed, add the sausage slices to the dough and knead them in until they are evenly distributed.

After a 30-minute rest period, divide the dough and again allow it to recover for 10 to 15 minutes. Form the dough and place it into greased round, covered molds with lids. Allow the dough to proof until the molds are 2/3 filled, then close the lids and bake as directed. After baking, allow the bread to cool in the molds for several minutes, then remove it and allow it to cool or temper completely.

Potato Bread

Allow 750 g (26.46 oz) peeled, starchy potatoes for each 1.5 kg of flour. Steam the potatoes, mash them well, then set them aside to cool.

Dissolve the yeast in part of the milk and water mixture and allow it to rest for several minutes (Exhibit 11–31). Add the rest of the liquid, the flour, salt, and mashed potatoes. Mix together the ingredients to form a *pâte bâtarde*. Sprinkle with a little of the additional milk and water mixture if necessary. Form into elongated dough loaves, and be sure to allow the finished bread to cool or temper on grates or screens after baking.

Exhibit 11-31 Potato Bread

Formula

Wheat flour	2.0 kg (70.55 oz)	100%
Mashed potatoes	1000 g (35.27 oz)	50%
Yeast	67 g (2.36 oz)	3.33%
Salt	54 g (1.90 oz)	2.66%
Water + Milk (50/50)	266 g (9.38 oz)*	13.33%
TOTAL DOUGH WEIGHT	3387 g (119.47 oz)	

Procedures

Mixing: 1st speed	3 min
Beater: 2nd speed	6 min
Dough temperature	26° C / 78.8° F
1st fermentation	75 min
Punch down after	35 min
Dividing & forming	15 min
2nd fermentation	45 min
Baking (225° C)	35 min
PRODUCTION TIME	3 h

* Add milk and water as necessary to achieve a dough of desired consistency.

Soya (Soy) Bread

Soya bread might, from a certain viewpoint, seem to be better balanced nutritionally than conventional breads. The addition of 15% soya (based on flour weight), either in the form of flour or coarse semolina, effectively raises the protein level of the bread. It also provides a better amino acid equilibrium and compensates for the lysine deficiency of wheat proteins, since soya has 10 times more of this amino acid than wheat. On the other hand, it notably diminishes the overall quality of the bread, causing decreased volume and undesirable taste.

Since the addition of soya lessens the physical strength of the dough, the flour used for this product should be fairly strong. If the flour is of only average baking quality, it should be reinforced with a little powdered wheat gluten.

Soya may be added either in the form of soya flour or as coarse soy semolina. The semolina adds color to the bread crumb, making it more attractive. In either case, the soy products are produced from soybeans which have been enzyme-deactivated by heat treatment, defatted, dried, and then milled into soya flour or semolina.

Either the traditional straight dough (Exhibit 11–32) or the prefermented dough (Exhibit 11–33) methods

Exhibit 11–32 High Protein Soya Bread: Straight Dough Method

Formula

Wheat flour	8.35 kg (294.53 oz)	
Powdered wheat gluten	0.15 kg (5.29 oz)	
Soya flour	1.0 kg (35.27 oz)	
Soya semolina (1)	500 g (17.64 oz)	
Total flour	10.0 kg (352.73 oz)	100%
Salt	200 g (7.05 oz)	2%
Yeast	250 g (8.82 oz)	2.5%
Ascorbic acid	(400 mg)	(40 ppm)
Water	7.2 l (253.97 oz)	72%
TOTAL DOUGH WEIGHT	17.65 kg (622.57 oz)	

Procedures

Mixing: 1st speed	4 min
(Oblique axis): 2nd speed	12 min
Dough temperature	25° C / 77° F
1st fermentation	70 min
Divide, rest	25 min
Dough forming	
2nd fermentation	65 min
Baking 230° C	30–35 min
PRODUCTION TIME	3 h 30 min

Exhibit 11–33 High Protein Soya Bread by the Prefermented Dough Method

Formula

	Weight		Prefermented Dough	Final Dough
Wheat flour	8.35 kg (294.53 oz)		2.0 kg (70.55 oz)	6.35 kg (223.99 oz)
Powdered gluten	0.15 kg (5.29 oz)			150 g (5.21 oz)
Soya flour	1.0 kg (35.27 oz)			1.0 kg (35.27 oz)
Soy semolina (1)	0.5 kg (17.63 oz)			500 g (17.64 oz)
Total flour	10.0 kg (352.73 oz)	100%		8.0 kg (282.19 oz)
Salt	200 g (7.05 oz)	2%	40 g (1.41 oz)	160 g (5.64 oz)
Yeast	250 g (8.82 oz)	2.5%	40 g (1.41 oz)	210 g (7.41 oz)
Ascorbic acid	(350 mg)	(35 ppm)		(350 mg)
Water	7.2 l (253.99 oz)	72%	1.36 kg (47.97 oz)	5.84 kg (206 oz)
Preferment dough			3.44 kg (121.34 oz)	3.44 kg (121.34 oz)
TOTAL DOUGH WEIGHT				17.65 kg (62.26 oz)

Formula

Mixing: Low speed	4 min	4 min
(Oblique axis): Second speed	10 min	10 min
Dough temperature from mixer	24° C / 75° F	25° C / 77° F
Prefermented dough rest time	4 h at 24° C / 75° F or 16 h at +4° C / 39° F	
First fermentation		35 min
Division, rest, forming		25 min
Second fermentation		65 min
Baking @ 230° C		30–35 min
PRODUCTION TIME		2 h 55 min

may be used to advantage in soya bread production. To obtain the best results, consider the following:

- The amount of ascorbic acid used should be relatively high.
- Dough mixing should not be excessive, so that normal dough consistency can be maintained.
- The first fermentation should be average in length, while the second should be slightly reduced.
- Baking should be done at a rather low temperature, and bake time should be slightly longer.

NOTES

1. The order of the recipes in this chapter is slightly different from the French text. This was done in order to reflect the relative importance of the products in the French marketplace, since the product order in the French text might tend to give the North American reader an inaccurate concept of the importance of such variety products in comparison to more commonly encountered breads.
2. *Farine de gruau* is a relatively strong flour that was formerly stone milled from the fine outer particles of the wheat berry. This has been replaced in more recent times by the milling of special strong, high protein flours. Although this would at first glance seem to correspond to an American patent flour, in actual practice this is not the case, and patent flour from most American wheat varieties are simply not suitable for French breads. For a more complete discussion of this topic, see the articles which discuss Professor Calvel's work with American flours at Kansas State University. In actual practice, the relative strength of North American bread flours is such that those appropriate for the other recipes in the book will also work for *pain de gruau*.
3. Again, standard North American bread flour is strong enough to be used in place of *farine de gruau* in virtually all instances.
4. In the United States, the standard for rye bread is given in the *Code of Federal Regulations*, Chapter 21.
5. Enzymes are an important factor in rye bread doughs because they remain active for a time following starch gelatinization during baking and break down the starches. This is a desirable effect as long as it is kept at a reasonable level by the inhibiting effect of sourdough or other acidic conditions, and enzymatic action helps to keep the crumb of well-made rye bread tender and relatively moist. Many North American rye flours are low in enzyme activity and can yield drier, tougher results. Although malt products are not traditional ingredients in rye doughs, they can be of benefit in correcting this problem.
6. Overmixing would break down the relatively small amount of gluten supplied by the wheat flour in the recipe and would adversely affect the volume and structure of the loaf.
7. In the United States, one even hears occasionally of 12-grain breads. Many of these additional "grains," however, are actually seeds or legumes, including such things as tef, quinoa, buckwheat, millet, and lentils.
8. Some technical authors refer to this method as the "four-part" molding method. This "four part" or "multipart" type of molding is especially well suited for use with lidded baking pans, such as for Pullman loaves. It results in one of the better crumb structures among white pan breads and a loaf that hardly ever becomes deformed in baking.
9. Deformation of loaves in baking or at other production and packaging stages is commonly referred to as "crippling" in the United States, while the deformed loaves are called "cripples." The origins of this rather politically incorrect term are not known.
10. In U.S. and Canadian practice, bread pans are commonly coated with a commercially applied silicone polymer layer to provide excellent product release properties. Current law does not permit the application of these polymer coatings in the bread production facility. Bread pans are regularly sent to outside contractors to have the coatings chemically removed and reapplied. However, some bakers still use uncoated pans, which requires spraying of pans with an edible oil-based release agent at the beginning of each use cycle.
11. In the United States, the proportion and type of milk to be used in milk bread or rolls are closely regulated by the *Code of Federal Regulations*, Chapter 21.
12. The two sections must remain connected by a "hinge" of dough approximately 1/10 the thickness of the dough piece.
13. On a practical level, almost all normal American bread flours would be suitable for this purpose. Few bakers will be able to obtain Chopin alveograph test results.
14. In North American practice, many flours will meet production requirements. Even so, it is advisable to add from 2% to 3% gluten powder and a little ascorbic acid during mixing.
15. This is a good practice in any case, since the starch and water film deposited on mixing equipment are an excellent breeding ground for molds and bacteria of many types and may be the source of mold spore contamination of bakery products throughout the production area. Thorough sanitizing of equipment on a regularly scheduled basis is also required by law in many states and localities.
16. Cool pumpkin pulp and cooking water separately to obtain a finished dough temperature of 25° C / 77° F at the end of mixing.
17. Remember that any effort to add a significant amount of cinnamon will inhibit yeast growth, since cinnamon has some antimycotic properties.
18. Toulouse sausage is the basic French fresh sausage, made from medium ground pork seasoned with salt, pepper, and nutmeg and stuffed into hog casings approximately 1¼ inches in diameter. Italian or other similar sausage may be substituted.

Part V

Yeast-Raised Sweet Dough Products, Common and Dietetic Rusks, Breadsticks, Croissants, Parisian and Regional Brioches

Rusks and Specialty Toasted Breads

Chapter 12

- Rusks (Biscotte Courant)
- Gluten-free breads.
- Breadsticks and *grissini*.

RUSKS (BISCOTTE COURANT)

Rusks undergo two separate baking operations separated by an interval of 24 hours:[1] (1) production and baking of the loaves, followed by a rest or tempering period ranging from 18 to 24 hours, and (2) cutting the loaves into slices around 10 mm thick, which are then grilled and dried on a wire mesh belt that passes through a tunnel oven designed especially for this purpose. This double baking treatment is the source of the prefix "bis," (meaning "twice") in the [French] name of the product.

In France, rusks were first made in the cities by artisan bakers, from around the turn of the century until the mid-1930s. It was a very labor-intensive product, since the grilling or toasting was best done on baking sheets, which had to be preheated in order to achieve uniform coloration on both sides of the slices. This was not always done, and since there was also a lack of consistency in the grilling and the size and shape of the slices, there was a distinct lack of regularity and quality.

Moreover, the quality of rusks, as with all bakery products made from fermented dough, is strongly influenced by the quality of its basic raw materials, particularly by the quality of the flour. As in the case of white pan bread, the flour used should have a W value somewhere between 200 and 220, with a P/L ratio around 0.55. It is also important that the dry gluten content be equal to or above 10% by weight.

The characteristics of the fats used also must be taken into consideration. As a matter of preference, these should be anhydrous vegetable fats or margarines, with the point of fusion between 32 and 33°C in winter, and 35 to 36°C in summer.

Lecithin may be used for its emulsifying properties and the softening effect it exercises on gluten. Mono- and diglycerides of esterified fatty acids may also be used in the formula. Their emulsifying properties are even more pronounced than those of lecithin, and they are even more effective in facilitating the distribution of fat throughout the dough, a fine and even crumb structure, proper loaf volume, and the desired degree of crispness of the rusks after toasting.

Ascorbic acid is also used systematically in rusk production. It strengthens the dough, improves the mixing and handling tolerance of the *pâtons*, and thus improves the volume of the bread.

In general, rusk loaves are produced from a slightly firm dough. The dough has a high yeast content and may tolerate a relatively rapid production process because of the strengthening presence of ascorbic acid. Rusk loaves are generally baked in lidded pans or tins in tunnel ovens. Use of unlidded tins is still encountered, but rarely.

Like tinned sandwich breads, rusk loaves should have a fine, regular crumb structure. The forming of the *pâtons* has a major effect on achieving these qualities, and just as with tinned sandwich breads, several types of dough forming and panning techniques may be used. The most common methods are makeup into dough cylinders or "sausages" (the "cylinder" method) and the "twist" method of makeup into sectioned cylinders which are then turned 90 degrees and placed in a row in the pan. Some bakers make up the dough into a ball (the "ball" method), but the practice of twisting two smaller-diameter dough cylinders together before panning has largely been abandoned, even though it produces excellent results.

The three makeup methods in current usage—cylinder, twist, and ball—are all carried out automatically. In some cases, dough panning is automatic as

---

The recipes in this section, which are industrial in nature, have not been adjusted to North American flours.

well. However, dough panning must be done under a watchful eye, since the *pâtons* must frequently be placed into pans with careful human assistance.

Rusks for special dietary purposes or to assist in weight loss are also commonly produced. Formulas for the three most generally consumed types of rusks are given in Exhibit 12–1, and formulas for both low-calorie and high-calorie rusks are also provided in Exhibit 12–2 and Exhibit 12–3.

It may be useful to add a few additional instructions to go along with these formulas.

Flour Quality

Compared to the flour used for common rusks, flour used for a no-salt rusk should preferably have a stronger gluten and an alveogram P/L ratio between 0.6 and 0.7. For gluten rusks, it is preferable to use flour with a more extensible gluten, with the P/L ratio of the alveogram curve around 0.45. In addition, the vital wheat gluten powder added during mixing should also have good extensibility.

Mixing

The most commonly used type of mixer in rusk manufacture is the double-armed Artofex type, which is operated at a speed of around 40 rpm for approximately 20 minutes. Even though it is desirable to avoid mechanically overworking the dough, it should still be smooth at the end of mixing.

Dividing, Rounding, Rest Period, and Dough Molding

This phase of production is completely mechanical and is done with automatic equipment. In the production of cylinder or "sausage" shaped dough pieces, volumetric division is followed by rounding of the *pâtons*. These are then deposited into the pockets of a traveling overhead proofer, which delivers them to the automatic molder after an appropriate length of time.

Molders used in rusk manufacture generally have larger-diameter molding rollers than those used in standard breadmaking, and the *plage d'allongement* or rolling-out table is somewhat longer. Some molders are equipped with two pairs of upper rollers, one of which is mounted slightly higher than the other. The upper roller set impresses very lightly on the dough piece, but ensures that it will be fed exactly into the middle of the second roller pair. This allows a better and more gentle lamination of the dough piece. After molding, the dough pieces may be placed into pans automatically.

Exhibit 12–1 Formulas and Procedures for Common Types of Rusks

Formula

Ingredients	Common Rusk	No-Salt Rusk	High-Gluten Rusk
Flour	100.0 kg (3527.34 oz)	100.0 kg (3527.34 oz)	85.0 kg (2998.24 oz)
Gluten powder			15.0 kg (529.1 oz)
Salt	1.4 kg (49.38 oz)		1.0 kg (35.27 oz)
Yeast	5.5 kg (194 oz)	5.0 kg (176.37 oz)	4.0 kg (141.09 oz)
Sugar	4.5 kg (158.73 oz)	4.5 kg (158.73 oz)	1.2 kg (42.33 oz)
Malt	300 g (10.58 oz)	250 g (8.82 oz)	300 g (10.58 oz)
Fat	4.5 kg (158.73 oz)	4.5 kg (158.73 oz)	4.5 kg (158.73 oz)
Lecithin	150 g (5.29 oz)	150 g (5.29 oz)	150 g (5.29 oz)
Milk powder	1.5 kg (52.91 oz)	1.0 kg (35.27 oz)	1.0 kg (35.27 oz)
Ascorbic acid	(2000 mg [0.67 oz])	(3000 mg [0.105 oz])	(1000 mg)
Water	49 l (1728.4 oz)	48 l (1693.12 oz)	55 l
TOTAL DOUGH WEIGHT			

Procedures

Mixing: First speed	20 min	18 min	24 min
Dough temperature	(27° C)	(26° C)	(26° C)
Division / rounding	15 min	15 min	15 min
Rest	13–15 min	11–12 min	14–16 min
Dough forming	—		
Proofing (37° C)	50 min	45 min	45 min
Baking	30 min	30 min	30 min
PRODUCTION TIME	2 h 10 min	2 h	2 h 10 min

Rusks and Specialty Toasted Breads 133

Exhibit 12–2 Low-Calorie Dietary Rusk with Addition of Fermented Dough

Formula

	Weight		Fermented Dough	Final Dough
Wheat flour	65.0 kg (2292.77 oz)		12.0 kg (423.28 oz)	53.0 kg (1869.49 oz)
Fine bran	20.0 kg (705.47 oz)		3.5 kg (123.46 oz)	16.5 kg (582.0 oz)
Gluten	15.0 kg (529.1 oz)		4.5 kg (158.73 oz)	10.5 kg (370.37 oz)
Total flour	100.0 kg (3527.34 oz)	100%	20.0 kg (705.47 oz)	80.0 kg (2821.87 oz)
Water	75 l (2645.5 oz)		16.0 l (564.37 oz)	59.0 l (2081.13 oz)
Salt	1.4 kg (49.38 oz)	1.4%	280 g (9.88 oz)	1.12 kg (39.51 oz)
Yeast	4.0 kg (141.09 oz)	4%	800 g (28.22 oz)	3.2 kg (112.87 oz)
Sugar	1.5 kg (52.91 oz)	1.5%	300 g (10.58 oz)	1.2 kg (42.33 oz)
Lecithin	300 g (10.58 oz)	0.3%	60 g (2.12 oz)	240 g (8.47 oz)
Vegetable fat	500 g (17.64 oz)	0.5%	100 g (3.53 oz)	400 g (14.11 oz)
Ascorbic acid	(3000 mg [0.105 oz])	(30 ppm)	(600 mg [0.02 oz])	(2400 mg [0.08 oz])
FERMENTED DOUGH				37.54 kg (1324.16 oz) 37.54 kg (1324.16 oz)
TOTAL DOUGH WEIGHT				182.7 kg (6444.44 oz)

Procedures

	Fermented Dough	Final Dough
Mixing: Low speed	3 min	3 min
Mixing: Second speed	17 min	17 min
Dough temperature from mixer	(25° C)	(25° C)
First fermentation		3–4 h
Division, rest period		15 min
Dough molding		
Second fermentation / final proof		50 min
Baking		35 min
PRODUCTION TIME		2 h
TOTAL PRODUCTION TIME		5 to 6 h

Exhibit 12–3 High-Calorie Weight-Gain Rusk with Addition of Prefermented Dough

Formula

	Weight		Fermented Dough	Final Dough
Wheat flour	50.0 kg (1763.67 oz)	100%	10.0 kg (352.73 oz)	40.0 kg (1410.93 oz)
Salt	750 g (26.46 oz)	1.5%	150 g (5.29 oz)	600 g (21.16 oz)
Yeast	4.0 kg (141.09 oz)	8%	500 g (17.64 oz)	3.5 kg (123.46 oz)
Malt extract	3.5 kg (123.46 oz)	7%	700 g (24.04 oz)	2.8 kg (98.77 oz)
Sugar	4.0 kg (141.09 oz)	8%	500 g (17.64 oz)	3.5 kg (123.46 oz)
Lecithin	150 g (5.29 oz)	0.3%	15 g (0.53 oz)	135 g (4.76 oz)
Vegetable fat	6.0 kg (211.64 oz)	12%	450 g (15.87 oz)	5.55 kg (195.77 oz)
Liquid milk	14.5 kg (511.46 oz)	29%	2.9 kg (102.29 oz)	11.6 kg (409.17 oz)
Whey	1.0 kg (35.27 oz)	2.0%	200 g (7.05 oz)	800 g (28.22 oz)
Water	7 l (246.91 oz)	14.0%	1.4 l (49.38 oz)	5.6 l (197.53 oz)
Ascorbic acid	(100 mg)	(20 ppm)	(20 mg)	(80 mg)
FERMENTED DOUGH				16.815 kg (593.12 oz) 16.815 kg (593.12 oz)
TOTAL DOUGH WEIGHT				90.27 kg (3184.13 oz)

Procedures

	Fermented Dough	Final Dough
Mixing: Low speed	3 min	3 min
Mixing: Second speed	15 min	17 min
Dough temperature from mixer	(26° C)	(27° C)
Rest period in vat	60–90 min	5 min
Division, rounding, rest period		15 min
Dough molding		
Fermentation in proofer (39° C)		55 min
Baking (205° C)		32 min
Cooling & tempering		24 h
Toasting (195° C average)		11 to 30 min

In the "twist" method, the dough cylinder is cut into sections and each section is panned after being twisted 90 degrees. Sectioning, twisting, and pan loading all take place automatically. When the "ball" makeup method is used, the overhead proofer rest period is deleted. After mixing, the dough is fed into a multipocket automatic divider-rounder, and the dough pieces are automatically deposited into pans immediately after the rounding operation.

Proofing and Baking of the Pâtons

The automatic overhead proofer automatically feeds the tunnel oven, which is just after the proofer in the production line. The *pâtons* are generally baked in lidded pans. Baking time is a function of the size of the *pâton* and the temperature of the oven. Generally, baking time is around 25 minutes at a temperature of around 240°C (464°F).

Cooling and Tempering of Baked Loaves

After baking, the loaves are usually placed on racks equipped with very fine mesh screen trays. These racks are placed in tempering rooms where the loaves cool and "rest" for 18 to 24 hours. The atmosphere of the tempering rooms is cool and humid, so that the loaves undergo a slight hardening effect, and the internal humidity becomes evenly distributed throughout the loaf. This moisture redistribution facilitates slicing of the loaves and improves the regularity and evenness of toasting. During this tempering period the loaves must be protected from air currents in order to avoid harmful drying of outer crust layers.

Cutting and Sorting of Slices

Both the cutting and sorting operations may be done semiautomatically or on a completely automated equipment line.

Toasting and Drying

This two-part operation is carried out in a single pass through toasting ovens equipped with a traveling steel mesh belt. Total length of the toasting period is around 7 minutes at a temperature of 240°C (464°F). The baking temperature curve should be rather elevated during the first third of the oven passage, since maximum humidity loss is closely linked to maximum temperature. The humidity released from the product should be carefully retained in the first oven zone as much as possible, since it is important to obtaining even coloring of the slices. The rusks begin to color in the second oven zone, since they have already lost a large part of their moisture. For that reason it becomes necessary to draw off part of the moisture and gradually lower the oven temperature. In the third zone the water vapor is completely exhausted from the oven, and the rusks take on their final color while reaching the desired level of dryness.

After leaving the oven, the rusks are placed upright and sorted onto a traveling mesh belt for cooling. Finally, after about 20 minutes, they are wrapped and packaged for storage.

To obtain identical toast coloring on both faces, as well as to maintain the symmetry of the rusks, special care must be taken to maintaining the proper temperatures at the oven sole as well as in the oven vault throughout the toasting stage. When there is too much heat in the oven vault, the slices tend to become deformed, taking on a "roofing tile" shape. When there is too much heat on the oven sole, the "tile" arch occurs in the opposite direction.

Special attention is also necessary to achieve the proper degree of coloration, which should be a distinct golden-brown. *Without overdoing it, remember that the taste of rusks is directly correlated to the coloration.* A lack of proper color indicates a lack of dextrinization, which is detrimental to the taste of the rusk. This is especially encountered in the case of no-salt rusks, which generally lack flavor. For that reason, it is important to toast such products with special care in order to achieve the proper coloration. In addition, it is important to bear in mind that rusks have a general tendency to progressively lose their coloring over time when they are stored in clear protective packaging.

Packaging and Storage

After toasting and cooling, the rusks are wrapped by automatic packaging machines, either in clear plastic or cellophane film or in multilayer laminated film, which is impermeable to both light and humidity. The laminated films provide the best protection against these two damaging factors, especially when the films are lined with aluminum foil.

Generally, rusks that are wrapped in transparent film are for sale in large stores where the turnover of stock is rapid, and where the rusks hardly have the time to suffer from oxidation and from the possibility of rancidity that might result from their exposure to light.

Quality Aspects of a Good Rusk

The moisture content of a properly dried rusk is between 4% and 5%. A good rusk should exhibit a fine, regular, and homogeneous grain. The slice should be symmetrical, and the coloration should be uniform, with a rich golden color. It should be crisp and pleasant-tasting, without a tendency to break too easily.

Low-Calorie Rusk

This special dietary rusk has several positive aspects: it is low in sugars and high in proteins and crude fiber. It is a surprisingly agreeable product in terms of flavor, since the toasted bran it contains has a much more acceptable taste than it does when used in bread. It would thus seem to be well worth our attention.

On a practical production level, low-calorie rusks pose a problem in forming. It is difficult and time consuming to produce a cylinder or "sausage-shaped" dough piece, since the dough is very resistant to being lengthened. However, it can be formed into dough balls without any difficulties.

High-Calorie Rusk

Since the following formula is very rich, it requires some additional instructions and notable modifications to normal procedures. First of all, it should be noted that the malt to be used is a **nondiastatic** type.

The choice of added prefermented dough as a leavening method may be explained as follows:

- It almost allows the baker to dispense with the first fermentation period, which would be as much as 25 to 30 minutes with the straight dough method, and it simplifies the weighing and rolling out of the *pâtons* during dough forming. (It is possible to use a standard rusk dough provided that the proper adjustments are made to the formula.)
- It produces better results—there is better development of the *pâtons* in the oven, and the crumb structure is finer and more homogeneous, while the appearance of the loaves after baking is also superior.

On the other hand, the bulk fermentation period takes longer and is done at a temperature that is somewhat higher than normal.

As required by the high sugar content, baking takes place at a relatively low temperature (on average about 205°C or 400°F) for a longer period of time. For this reason, it is advisable that the *pâtons* be somewhat underproofed when they are placed in the oven.

The loaves should undergo a cooling and tempering period of at least 24 hours before toasting. Toasting is also done at a lower than average temperature (because of the high sugar content), on average about 195°C (384°F) for 11 to 12 minutes. When the required modifications to the basic formula and procedures are mastered, the high-calorie rusk that results has both an attractive appearance and a very rare and distinctive flavor profile.

Specialty Grilled Breads

Specialty grilled or toasted bread was initially developed during the 1960s. The product was first introduced by a single rusk manufacturer, but the example was rapidly followed by all of the other French rusk producers.

From the beginning, it was made both by the straight dough method and by the sponge and dough procedure. Those two methods are still in use today.

The dough formula used in the production of specialty toasted bread is less rich than that used in rusk manufacture (Exhibit 12–4). Both the sugar content and the fat content are lower. Also, the size of individual *pâtons* is much larger than for rusks. Makeup procedures, size and form of the baking pans, and baking procedures all differ from those used in rusk production.

Grilled or pretoasted bread is made from large dough pieces that undergo a rest period in an overhead proofing chamber after rounding. Then they are formed into large sausage shapes and deposited into relatively wide rectangular molds. They are then lightly flattened either by pressure or by lengthwise sheeting so that the individual dough piece entirely covers the surface of the mold. The special mold used for this product is a shallow baking tray, with sides that are about 4 to 5 cm high.[2]

The bread baked in these special open-topped tins is from 15 to 18 cm wide and around 40 cm long. At its highest point, the dome-shaped top is about 10 to 12 cm above the edge of the tin.

After baking, the loaves cool and temper for a minimum of 24 hours before being sliced. The large slices are toasted on traveling steel mesh belts, which pass through special tunnel ovens reserved for this purpose. Toasting is somewhat more complicated than for ordinary rusks. Because of their shape, the slices cover the surface of the mesh belt rather unevenly. This allows the passage of heat through the openings between the slices and often results in excessive coloration and overtoasting of the edges of the slices. For that reason, the production of specialty pretoasted bread requires very careful attention.

The preferred structure of the crumb is a little less regular than that of common rusks. Specialty toast

Exhibit 12–4 Specialty Bread for Pre-Toasting, Made with Addition of Fermented Dough

Formula

	Weight		Fermented Dough	Final Dough
Wheat flour	60.0 kg (2116.40 oz)	100%	12.0 kg (423.28 oz)	48.0 kg (1693.12 oz)
Salt	1.2 kg (42.33 oz)	2%	240 g (8.47 oz)	960 g (33.86 oz)
Yeast	2.7 kg (95.24 oz)	4.5%	540 g (19.05 oz)	2.16 kg (76.19 oz)
Malt extract	600 g (21.16 oz)	1%	120 g (4.23 oz)	480 g (16.93 oz)
Sugar	1.5 kg (52.91 oz)	2.5%	300 g (10.58 oz)	1.2 kg (42.33 oz)
Vegetable fat	1.8 kg (52.91 oz)	3%	360 g (12.7 oz)	1.44 kg (50.79 oz)
Milk powder	600 g (21.16 oz)	1%	120 g (4.23 oz)	480 g (16.93 oz)
Ascorbic acid	(1200 mg [0.04 oz])	(20 ppm)	(240 mg)	(960 mg [0.03 oz])
Water	31.2 l (1100.53 oz)	52.0%	6.24 l (220.1 oz)	24.96 l (880.4)*
Fermented dough				19.92 kg (702.65 oz) 19.92 kg (702.65 oz)
TOTAL DOUGH WEIGHT				99.60 kg (3513.23 oz)

Procedures

		Fermented Dough	Final Dough
Mixing: Low speed		3 min	3 min
Mixing: Second speed		13 min	13 min
Dough temperature from mixer		(26° C)	(26° C)
First fermentation		75–90 min	5 min
Division, rest period			15 min
Dough molding, punching down			10 min
Fermentation in baking tin (37° C)			45 min
Baking (240° C average)			25 min
Production time			1 h 56 min
TOTAL PRODUCTION TIME			3 h 11 min to 3 h 26 min

*There seems to be a mistake in the original table which gave 6.24 l in both columns.

---

should also be slightly less crisp and a little less sweet than a rusk. It should be noted that when the dough is partly leavened by the sponge and dough method, the resulting toast is a little more crisp and "melts" in the mouth less readily than toast made by the straight dough method. When the sponge and dough product is dipped in soup or coffee, it becomes soaked more slowly and holds together better.

Holland Rusks: The Original Grilled Toast

Grilled toast, which was first produced in Holland, is both little known and rarely made in France. However, it is a distinctive product that deserves to be better known. It is very rich, very light, and very crisp. Grilled toast is made from strong flours (with a W value around 280) and a protein level of at least 12% (Exhibit 12–5). The formula includes a high proportion of sugars (glucose and sucrose) and in Holland also contains a whole range of emulsifiers, ranging from lecithin to mono- and diglycerides of fatty acid esters and even a gel that contains a small amount of Marseilles soap and a little potash. When grilled toast is made in France, the use of the gel is not permitted.

Dough mixing involves at least two distinct steps: in the first, the baker mixes two thirds of the flour and the desired quantity of water, yeast, and glucose. When the dough has become fairly homogeneous, the mixer is stopped and a 15-minute rest period is begun, which is a sufficient time for the yeast to become fully active. When this period has elapsed, the second step begins as sucrose, salt, fat, eggs, emulsifiers, ascorbic acid, and the remaining flour are added to the mixing bowl, and mixing continues until the dough has become very smooth.

The first rest period in the bowl may last between 15 and 20 minutes, if the dough pieces are divided with a semiautomatic divider-rounder, or may be almost entirely eliminated if these operations are completely automated. Furthermore, in semiautomatic operation, there should be a preliminary weighing of the large dough pieces, followed by a first rounding.

In the first instance, the large dough pieces that are to be cut up should be weighed. After dividing and preliminary rounding, they are left to rest "on the bench" for 10 minutes before being run through the divider-rounder. In the second case, the automatic divider-rounder is fed with dough continuously.

Exhibit 12–5 Formula and Procedures for Holland Rusks

Formula

Flour	40 kg (1410.93 oz)	100%
Salt	480 g (16.93 oz)	1.2%
Yeast	3 kg (105.82 oz)	7.5%
Eggs	3.2 kg (112.87 oz)	8%
Milk powder	800 g (28.22 oz)	2%
Fat	3 kg (105.82 oz)	7.5%
Crystalline glucose	6 kg (211.64 oz)	15%
Sugar	2 kg (70.55 oz)	5%
Lecithin	120 g (4.23 oz)	0.3%
Ascorbic acid	(1200 mg [0.04 oz])	(30 ppm)
Water	17.6 l (620.8 oz)	44%

Procedures

Mixing: 1st speed	5 min
Rest period	15 min
Mixing: 2nd speed	15 min
Dough temperature	(27° C)
Rest in bowl	10 min
Divide, round	20 min
Rest, punch down	10 min
Proofing chamber	40 min
Baking	7 min
Cool / temper	5–6 h
Slice / toast	7–8 min
TOTAL TIME	7 h 9 min to 8 h 10 min

After rounding, the dough balls are deposited onto individual metal trays or onto a lightly greased traveling metallic band. Following another 10-minute rest period, the dough balls are lightly degassed and placed on baking trays; each ball is covered with an individual baking mold which is used especially for this product. The mold is a cover that measures 100 mm in diameter, with an edge between 28 and 30 mm tall. The upper part of the mold is pierced with a number of regularly spaced small holes.

The loaded molds are next passed through a proofing chamber or tunnel. When the dough has proofed to the level of the second or third perforation in the mold, the molds and their contents are loaded into the baking chamber. Baking is carried out relatively rapidly and at a fairly high temperature—around 240°C for only about 7 minutes.

The small loaves are depanned as soon as they are removed from the oven. Since the crust is rather thin, they are initially cooled under controlled conditions to reduce the risk of excessive crust shrinkage and wrinkling. After tempering for 4 to 6 hours, the loaves are sliced in half horizontally before toasting. The halves are toasted with the cut side up at an average temperature of 210°C.

As in the case of conventional rusks, Holland rusks should be very dry before packaging. As in the case of all high-fat products, this specialty toast must be carefully protected against both humidity and the effects of exposure to light. When correctly formulated and produced, this is a very high-quality bakery food.

The high-production lines on which these specialty toasts (rusks) are made today are completely automatic. In addition to the traditional Holland rusk, a very good variation may be produced by incorporating around 30% raisins (on flour weight) into the dough at the end of mixing.

Zwieback

Zwieback is a German product that is little known in France. As the name indicates [zwie-two], it is somewhat related to rusks in that it is baked two times. It differs from common rusks in that it is richer and toasted differently. It is largely produced industrially but is also made at an artisan level.

As an artisan product, Zwieback may be sold either fresh or toasted. It is made by aligning small oval dough balls of yeast-raised sweet dough about 8 to 10 mm from one another. After proofing and baking, the small rolls are joined together like Siamese twins, and the baking sheet is covered with a network of interconnected rolls.

The purchaser may buy these fresh rolls in linked groups of varying sizes. If any remain unsold, they are sliced and toasted the next day. However, they are toasted to a much greater degree than common rusks, since they are toasted through and rather dark brown.

At the industrial level, production is almost completely automatic. The sponge and dough system of leavening is generally used (Exhibit 12–6), and the dough is divided with multipocket divider-rounders. Loading of the tall, rectangular-section pans is automatic. Toasting is done after a cooling and tempering period of at least 24 hours, as in the case of the *biscotte*[3] in a relatively cool oven (around 210°C). As in the case of artisan-produced zwieback, the slices are toasted through to an identical color.

As previously mentioned, zwieback is a rich product eaten primarily by infants or by persons who must consume a large number of calories.

GLUTEN-FREE BREADS

The starches shown in Exhibit 12–7 may be included as a base mixture for this formula.

This no-gluten bread is a special dietary product, produced especially for those individuals who suffer

Exhibit 12–6 Zwieback, Made by the Sponge and Dough Method

Formula

	Weight		Fermented Dough	Final Dough
Wheat flour	30.0 kg (1058.2 oz)	100%	5.0 kg (176.37 oz)	25.0 kg (881.83 oz)
Yeast	2.1 kg (74.07 oz)	7%	100 g (3.53 oz)	2.0 kg (70.55 oz)
Salt	420 g (14.81 oz)	1.4%		420 g (14.81 oz)
Malt	150 g (5.29 oz)	0.5%		150 g (5.29 oz)
Lecithin	90 g (3.17 oz)	0.3%		90 g (3.17 oz)
Milk powder	780 g (27.51 oz)	2.5%		780 g (27.51 oz)
Sugar	4.8 kg (169.31 oz)	16%		4.8 kg (169.31 oz)
Fat	2.4 kg (89.66 oz)	8%		2.4 kg (89.66 oz)
Ascorbic acid	(1050 mg [0.037 oz])	(35 ppm)		(1050 mg [0.037 oz])
Water	15.0 l (529.10 oz)	50%	2.85 kg (100.53 oz)	12.25 kg (432.1 oz)
SPONGE			7.95 kg (280.42 oz)	7.95 kg (280.42 oz)
TOTAL DOUGH WEIGHT				55.84 kg (1969.66 oz)

Procedures

Mixing: Low speed	8 min	3 min
Mixing: Second speed		15 min
Dough temperature from mixer	(26° C)	(26° C)
Rest (in bowl or container)	4 min	
First fermentation		10 min
Division, rounding		20 min
Mold loading		
2nd fermentation in baking tin		50 min
Baking (225° C average)		20 min
Cooling & tempering		24 h
Toasting (210 ° C)		12 min

Exhibit 12–7 Gluten-Free Bread

Formula

Ingredients	Weight
Wheat starch	400 g (14.11 oz)
Corn starch	280 g (9.88 oz)
Pregelatinized starch	120 g (4.23 oz)
Dried potato flakes, hydrated with 600 g of boiling water	100 g (3.53 oz)
Soy flour (deodorized and heat-treated)	100 g (3.53 oz)
Total starch component	1.0 kg (35.27 oz)
To the above base mixture, the baker should add:	
Salt	20 g (0.71 oz)
Yeast	50 g (1.76 oz)
Dextrose (glucose)	30 g (1.06 oz)
Vegetable fat	30 g (1.06 oz)
Milk protein (substitute: non-fat milk powder)	40 g (1.41 oz)
Water (to be added to the mixture– does not include the water used to hydrate the dehydrated potato flakes)	500 g (17.64 oz)

Procedures

Mixing: First speed	12 min
Dough temperature from mixer	(26° C)
Primary fermentation	20 min
(The dough will be soft at the end of mixing, but will become more firm during the rest/fermentation period)	
Division, rounding, dough forming (dust lightly with starch)	10 min
Secondary fermentation (Dough pieces are deposited into pans)	20 min
Baking (220° C / 428° F)*	35–40 min

*To prevent the loaves from bursting, oil the dough pieces and cover them with aluminum foil for the first 20 min of baking, then remove the foil.

from an intestinal disorder called celiac disease. This illness requires that the patient observe a very strict diet that prohibits any consumption of gliadin, the water-soluble component of gluten. Although the disease is not widespread, nearly one person in a thousand is affected. Wheat proteins are not the only ones forbidden to these individuals, but those of rye, barley, and oats as well.

Fortunately, they are able to assimilate cereal starches and starches from various vegetable origins without problems. To make a no-gluten bread to meet their needs, the baker may develop starch-based formulas that may also contain flours from corn (maize), rice, millet, or buckwheat.

All of these ingredients have poor plastic properties and result in porous doughs that are generally unable to retain fermentation gases. For that reason it is necessary to add binding ingredients, which may include eggs, gelatinized starches, and milk proteins. To improve the retention of fermentation gases, the baker may also add other natural substances derived from algae or carob beans.

BREADSTICKS AND GRISSINI

These two bakery products are made from very similar formulas (Exhibit 12–8). Artisan bakers have made them by hand for quite a long time: breadsticks on special baking sheets stamped with elongated molds in series, and *grissini* on flat baking sheets. The latter are an Italian specialty, especially from the region of Turin, where one may still find *grissini* made by hand and baked directly on the oven sole. Today, however, these two products are generally produced by industrial bakeries using large-scale equipment.

In the majority of cases, production begins with a rather stiff dough that is fermented for a very brief period before being sheeted and laminated several times. The dough sheet is then passed between two rollers with regularly spaced grooves around the circumference. Somewhat like a pasta-forming die, these grooves continuously force the dough into "ropes," which are cut to the desired length (from 15 to 25 cm) and deposited into special gang molds where they are aligned as necessary, proofed, and baked without undergoing any deformation.

The dough is usually leavened by the sponge and dough method, but the straight dough method is used as well. The dough is baked in tunnel ovens at temperatures between 230 and 240°C (464°F), with a great deal of injected steam. After baking, breadsticks are from 20 to 25 mm in diameter, while the *grissini* are around 10 mm thick. Breadsticks commonly go through a drying chamber or tunnel after baking.

In addition to conventional products of this type, there are also special dietary breadsticks and *grissini*

Exhibit 12–8 Small-Scale Formulas for Breadsticks and *Grissini*

Formula

Ingredient	Breadsticks		Grissini	
High-grade flour	5.0 kg (176.37 oz)	100%	5.0 kg (176.37 oz)	25.0 kg (881.83 oz)
Water	2.25 kg (79.37 oz)	45%	2.25 kg (79.37 oz)	45%
Salt	100 g (3.53 oz)	2%	90 g (3.17 oz)	1.8%
Yeast	175 g (6.17 oz)	3.5%	200 g (7.05 oz)	4%
Malt	75 g (2.65 oz)	1.5%	75 g (2.65 oz)	1.5%
Sugar	200 g (7.05 oz)	4%	125 g (4.41 oz)	2.5%
Fat	375 g (13.23 oz)	7.5%	250 g (8.82 oz)	5%
Milk powder	75 g (2.65 oz)	1.5%	75 g (2.65 oz)	1.5%
Ascorbic acid	(50 mg)	(10 ppm)	(50 mg)	(10 ppm)

Procedures

Mixing: Low speed	3 min	3 min
Beater: Second speed	8 min	8 min
Dough temperature from mixer	(25° C)	(25° C)
First fermentation	50 min	40 min
Division, rest period	25 min	30 min
Dough forming		
2nd fermentation	70 min	65 min
Baking (230° C average)	15 min	12 min
PRODUCTION TIME	2 h 51 min	2 h 38 min

made without added salt or from whole grain or gluten-enriched flours.

Exhibit 12–8 shows formulas for both products, along with a set of procedures that should allow the baker to make them on a small artisan scale.

Turin-Style *Grissini*

Grissini may also be made without any real dough forming (Exhibit 12–9). When using this method, the baker cuts narrow sticks from a small rectangle of dough. These sticks are then rolled and stretched to length before being lined up on ordinary flat baking sheets, or as in Turin, stretched to length and then rolled in semolina flour. After a short rest period of around 30 minutes on cloth, the sticks are deposited onto a wide baker's peel and placed directly on the oven sole to bake.

After mixing, the dough is rolled and molded, and degassed into the form of a long cylinder (sausage) from 15 to 20 cm long. It is left to rest on a floured bench for 90 minutes. During this entire period it is covered and carefully protected from drying and crusting so as not to restrict its expansion.

At the end of this rest period, dough sticks are cut transversely from the dough cylinder and rolled and stretched from 1.5 to 2.0 m in length. They are placed either on the belt of the oven loader or on a baker's peel, using fine rice semolina as dusting material. The formed *grissini* are then placed directly on the oven sole for baking for 7 to 8 minutes. They are fairly well dried in baking, but the baker must avoid overbaking, which would result in actually toasting the product all the way through.

After cooling, the long *grissini* are broken into pieces around 25 cm long and sold loose in bulk or packaged into small bags.

Exhibit 12–9 Formula and Procedures for Turin-Style *Grissini* from Straight Dough

Formula

Flour	3 kg (105.82 oz)	100%
Salt	45 g (1.59 oz)	1.5%
Yeast	45 g (1.59 oz)	1.5%
Non-diastatic malt	120 g (4.23 oz)	4%
Fat	120 g (4.23 oz)	4%
Water	1.8 l (63.49 oz)	60%

Procedures

Mixing: 1st speed	6 min
Beater: 2nd speed	5 min
Dough temperature	(22° C)
1st fermentation	90 min
Cutting, forming	6 min
Baking (235° C)	7–8 min
TOTAL TIME	1 h 55 min

NOTES

1. Rusk production is similar to the production of Italian **biscotti** in that it involves two distinct baking steps.
2. Somewhat like the pan commonly used for home baking of brownies in the United States.
3. The original French text stated "as in the case of the baguette" but this is certainly a typographical error. What Calvel means to say is *la biscotte*.

Yeast-Raised Sweet Doughs

Chapter 13

- Traditional croissants.
- Chocolate-filled buns from croissant dough.
- Snail rolls.
- Brioches.

TRADITIONAL CROISSANTS

The story of the origin of the croissant is one that certainly bears repeating. It appears to have first been developed in Vienna, Austria, in 1683. The Austro-Hungarian Empire was at war with the Turkish Empire, the Turkish army had laid siege to Vienna, and the siege had been dragging on for some time. To bring it to a close, the Turkish soldiers began to dig a tunnel beneath the fortifications with the intention of taking the city's defenders from the rear. What the Turks did not realize is that they were digging toward the basement of a bakery, and that the noise of their digging would alarm the bakers who were working there.[1] The bakers sounded the alarm, and the Turks were the ones who were surprised, resulting in the defeat of their army.

The grateful emperor of Austria-Hungary awarded a number of privileges to the brave bakers of Vienna. In thanks to the Emperor and in celebration of the victory, the bakers created a small roll in yeast-leavened sweet dough. The roll was made in the form of the crescent moon—the emblem that still appears on the Turkish flag—and the *croissant* was born!

A century later, following the marriage of the Austrian princess Marie-Antoinette to the future French King Louis XVI, the croissant began to be made at the Court of Versailles. It was successful there, and little by little it was adopted by Paris and then by the rest of France.

However, the great surge in the popularity of the croissant dates only from the World's Fair of 1889, during which croissants were again made by Viennese workers. It was from this period that the small milk rolls, brioches, and croissants made from yeast-leavened sweet dough became known as Vienna goods. In fact, during the following three or four decades the Paris bakers who make these types of products were known as "Vienna bakers."

From its origins until about 1920, the so-called Vienna croissant was made from an unlaminated, nonflaky, yeast-raised sweet dough. Around that time, French bakers developed and began to regularly produce the laminated, flaky croissant as we know it today. The laminated-dough flaky croissant is thus of French origin, and the fermentation of the dough, which contributes to the formation of its flaky layers, represents considerable progress: as a result, the croissants are more voluminous, lighter, more delicate, and more flavorful.

Croissant Production

It is important to always use the proper quality and types of materials when making croissants. Choose a pure wheat flour (not mixed grain) with strength and baking qualities slightly superior to ordinary bread flour: the W value should be at least 220, with a P/L ratio around 0.6, since these qualities result in extensible doughs with good shape retention. The enzymatic activity should be rather weak, with a falling number value equal to or greater than 250 seconds.

Fats used for croissants—whether butter or margarine—should have good plasticity and stability. The moisture content of the butter should be 15% at most, and the point of fusion of margarine should not be higher than 36°C.

Malt extracts or syrups are preferable to malt flours.

The mixing times in this chapter are for planetary mixers.

Formula and Procedures

For each kilogram (35.27 oz) of flour, the basic formula generally includes 20 to 24 g (0.78 oz) salt, 30 to 40 g (1.23 oz) fresh yeast, 80 to 120 g (3.53 oz) sugar, 20 g (0.71 oz) milk powder, 30 to 40 g (1.23 oz) fat, 10 g (0.35 oz) malt extract or malt syrup, 540 g (19 oz) water, one egg (sometimes), and 430 g (15.17 oz) butter or margarine for laminating (Exhibit 13–1 and 13–2).

The addition of 3% to 4% fat to the dough at mixing improves the extensibility of the dough and simplifies lamination and dough forming. In the same way, the addition of approximately one egg per kilogram of flour (5% on flour weight) improves both the plastic properties of the dough and the taste of the croissant.

The type of fat used in dough lamination may make some changes necessary in the amount of salt and sugar used in the base formula. For example, when using sweet, unsalted butter, it is advisable to add formula salt at the rate of 24 g per kg of flour. When using margarine, which is lightly salted, the amount of salt used in the formula is only 20 g per kilogram of flour.

When sugar is used with butter, use only 80 to 100 g per kilogram of flour to avoid interference with the aromas carried by the butter. When using margarine, sugar should be used at a rate of 100 to 120 g per kilogram of flour to improve the taste of the product.

Dough Mixing

In the early days of flaky, laminated croissant production, the dough was given a first fermentation as long as 6 to 7 hours. This long fermentation accomplished part of the dough development, which was then completed by the lamination process. Mixing hardly went beyond a good light blending of ingredients.

Today the first fermentation takes place much differently, and the dough is often refrigerated or retarded. Because of this change, and because bakers add a small amount of ascorbic acid, the level of

---

French margarine is salted, so in Exhibit 13–2, Professor Calvel has reduced the salt to 2% (40g; 1.41 oz) and increased the sugar from 9% (180g; 6.35 oz) to 11% (220g; 7.76 oz). Professor Calvel has written that the autolysis is advisable for all croissant dough, and this recipe may be adapted for the use of butter by adjusting the quantities of salt and sugar.

Exhibit 13–1 Formula and Procedure for Traditional Butter Croissants

Butter Formula

Flour	2 kg (70.55 oz)	100%
Water	1.2 l (42.33 oz)	60%
Salt	48 g (1.69 oz)	2.4%
Yeast	70 g (2.47 oz)	3.5%
Sugar	180 g (6.35 oz)	9%
Malt extract	20 g (0.71 oz)	1%
Milk powder	30 g (1.06 oz)	1.5%
Butter	40 g (1.41 oz)	2%
Ascorbic Acid	(40 mg)	(20 ppm)
TOTAL DOUGH	3.588 kg (126.56 oz)	

Procedures

Mixing: 1st speed	4 min
Beater: 2nd speed	4 min
Dough temperature	(22° C / 72° F)
1st fermentation	90 min @ 24° C / 75° F
	90 min @ +4° C / 39° F
Lamination	35 min
Divide, form	25 min
2nd fermentation	75 min @ 27° C / 81° F
Baking 230° C / (445° F)	20 min
TOTAL TIME	5 h 55 min

Butter used in lamination: 3.488 x 0.25 = 870 g

Exhibit 13–2 Formula and Procedure for Croissants Made with Margarine by Autolysis Rest Method

Margarine Formula

Flour	2 kg (70.55 oz)	100%
Water	1.2 l (42.33 oz)	60%
Salt	40 g (1.41 oz)	2%
Yeast	70 g (2.47 oz)	3.5%
Sugar	220 g (7.76 oz)	11%
Malt extract	20 g (0.71 oz)	1%
Milk powder	30 g (1.06 oz)	1.5%
Margarine	40 g (1.41 oz)	2%
Ascorbic acid	(40 mg)	(20 ppm)
TOTAL DOUGH	3.576 kg (126.14 oz)	

Procedures

Mixing: 1st speed	4 min
Beater: 2nd speed	6 min
Autolysis rest	30 min
Dough temperature	(22° C)
1st fermentation	60 min @ room temperature
Refrigeration	15 h @ +4° C
Lamination	35 min
2nd fermentation	80 min @ 33° C
Baking (230° C / (445° F)	20 min
TOTAL TIME	19 h 18 min

Margarine used in lamination: 3.540 x 0.25 = 885 g

> French butter is churned from cream that has fermented (in the same manner as crème fraiche or yogurt) and it has a complex, nutty flavor. In addition, great care is taken to eliminate excess moisture, therefore producing a firmer texture, ideal for croissant, brioche, and puff pastry production.
>
> More important, still is that most French butter is unsalted. If salted butter must be used, reduce the salt quantity in the formula to 2% by flour weight (40 grams for the 2 kg recipes in this section). Salted butter contains approximately 2.5% salt. One kilogram of salted butter would add about 25 grams of salt to a recipe (and 25 grams less butter). It is therefore imperative that if salted butter is used, the quantity of salt in the recipe be readjusted.
>
> Those unused to working with butter must bear in mind that the low fusion point which makes butter disappear so deliciously on the tongue can be treacherous in warm climates. For proper lamination, the butter must be firm enough to roll evenly between each thin layer of dough, but not so cold that it becomes brittle and breaks into pieces. Bakers in many areas will find themselves refrigerating the buttered dough for longer than indicated. In the same way, proof box temperatures should not be too warm because the butter can liquify and leak out of the croissants as they rise.

---

primary fermentation has been greatly reduced, and the amount of dough development formerly achieved during this period has greatly diminished as well. It is necessary to take this into consideration and to slightly extend the duration of the mixing stage. In brief, the more the first fermentation period is reduced, the more mixing development of the dough becomes critical. At the present time, it is advisable to extend the mixing of croissant dough beyond the blending stage, taking care not to overmix the dough.

The Use of the Autolysis Rest Period

The autolysis rest period appears to have a number of very beneficial effects when used during croissant dough mixing. Autolysis permits a slight reduction in the duration of the mechanical work required in mixing the dough. It also increases the extensibility of the dough and aids both the lamination and the makeup of croissants. At the same time, it also yields croissants that are more regular in form and higher in volume. As noted previously, the autolysis rest period takes place between the blending of dough ingredients and actual mixing development.

The ingredients mixed in the blending stage include flour, water, sugar, milk powder, malt extract, butter, and occasionally some prefermented dough. These will first be mixed for around 4 minutes. The autolysis rest period that follows should be a minimum of 20 minutes, although 30 minutes is better.

When mixing is restarted, the following ingredients are added in order, one after another: yeast, salt, and ascorbic acid or some other additive. After mixing at first speed for one minute, the mixer is switched to second speed and mixing continues for 4 to 5 minutes maximum. The dough temperature at the end of mixing is normal, and the makeup method used is normal as well.

First Fermentation, Lamination, Makeup, and Proofing

Proper fermentation is essential to the production of a quality croissant, and it must be provided either by allowing plenty of time for the process or by improving the dough through the addition of a substantial amount of prefermented dough during mixing. In either case, the dough must contain a sufficient amount of organic acids *before* the laminating process begins.

As in the case of most traditional flaky doughs, the *tourage* or lamination process for croissants involves the enfolding and multiplication of thin layers of fat between two layers of dough (Figure 13-1). In the first stage, a thin sheet of cooled butter or margarine is deposited onto a rolled-out square of dough that has a surface area two times greater than that taken up by the thin fat layer. The thin fat sheet is covered or "imprisoned" by folding and slightly overlapping the sides of the dough sheet over the center fat layer so that it is impossible for the fat to escape when the dough is laminated or punched down. The amount of fat used is about or slightly greater than 250 g per kg of croissant dough (1 part butter to 4 parts dough by weight).

Laminating is improved by using both dough and fat that are at a fairly low temperature—about 12°C (54°F)—so that the thin sheets of butter or margarine stay relatively firm, while still remaining "plastic" in the form of continuous fat films between loaves of dough. Soft fat—either butter or margarine—tends to become sticky and makes the lamination process difficult, since it mixes with the dough and does not promote the baking-stage separation of the dough sheets between which the fat is enclosed. Refrigeration of the dough and a cool room temperature are practically indispensable to the production of a fine-quality flaky croissant.

The fine, crisp, multilayer structure of the croissant results from a folding technique or *tourage*, which uses either three simple folds or two double folds in

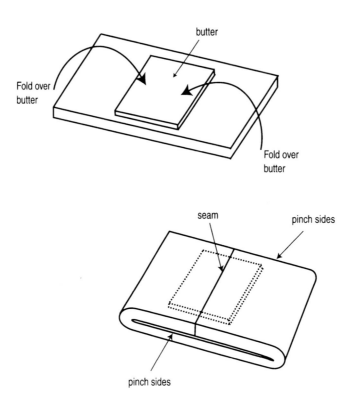

Figure 13–1 Lamination: Placing thin sheet of butter between layers of dough.

Cold butter is pounded with a rolling pin on the work surface until it is smooth and malleable, but remains cold. The dough is rolled to a rectangle and the butter placed on it (thickness of about 1/2 in). The flaps of dough on either side are then folded over the butter, forming a seam in the center of the dough, and the ends are pinched shut to enclose the butter completely.

the "pocketbook" or "book" style. In the first instance, after the fat layer is enclosed or "laminated" between two dough layers, the dough is rolled into a rectangle from 7 to 8 mm thick. The two outer thirds of the rectangle are then folded over the middle third, which forms the first folding or *tour*.

After a 10-minute rest period under refrigeration or very cool room conditions, the dough is again rolled out into a rectangle, and the two outer thirds are again folded over the middle third. This completes the second folding or second *tour*. This is followed by a second 10-minute cool rest period, after which the dough is again rolled out into a rectangle and again folded as shown in Figure 13–2. This is the third and last folding or *tour*.

In the second or "book" method (Figure 13–3), the fat layer is deposited between two dough sheets, and the dough is then rolled out into a rectangle between 6 and 7 mm thick. The two outside edges of the rectangle are folded to meet at the centerline of the rectangle. Then one half of the dough rectangle is folded over the other (much as the dust jacket of a book). The dough is then allowed to rest under cool conditions for 15 minutes, following which it is rolled out at a right angle from the original lamination, then folded again into a "book" form.

In both cases, the dough piece or pieces undergo a cool rest period following the completion of the folding or *tourage*. After 15 to 20 minutes, the dough pieces are rolled out to a thickness of about 3 mm. The dough sheet is then cut into triangles of 45 to 50 g in weight (Figure 13–4), and the individual triangles are rolled from the base to the point (Figure 13–5).

The individual dough rolls are then placed on a baking sheet, with the point of the triangle or "tongue" remaining visible on the top.[2] Those laminated with

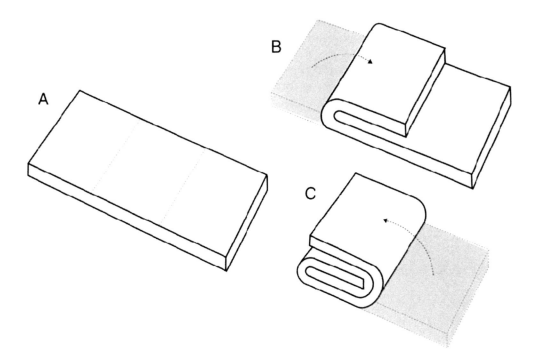

Figure 13–2 Three Fold Method. *Source:* Reprinted with permission from original drawing by Shelly Wohler. Produced for AIB course on croissant production.

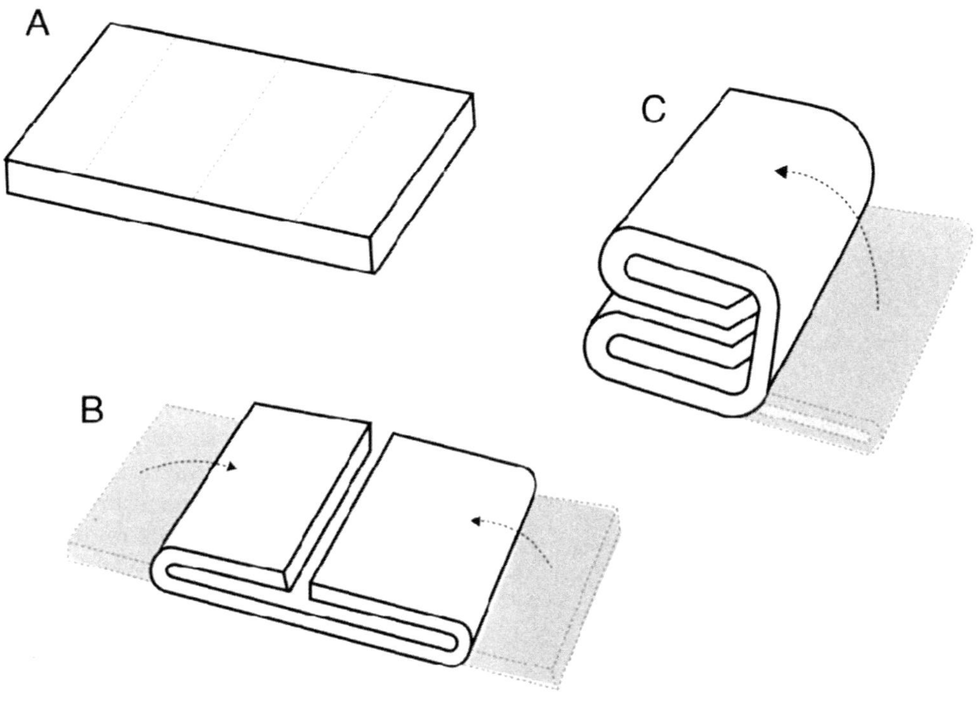

Figure 13–3 "Book" Method. The advantage of the "book" method is that an experienced individual, using a dough sheeter, can accomplish the 2 necessary book folds one after the other, with no rest period between. *Source:* Reprinted with permission from original drawing by Shelly Wohler. Produced for AIB course on croissant production.

margarine are formed into the crescent shape, while rolls laminated with butter are left in the straight form.

After a second fermentation at a maximum temperature for butter croissants of 28°C (82.4°F) or for margarine croissants at 32 to 33°C (89.6 to 91.4°F), the rolls are brushed with an egg wash and baked at around 230°C (445°F).[3]

The formula and procedures given here are suitable for producing croissants with margarine used as a laminating fat. As previously indicated, the salt level must be lowered to 2% (a rate of 20 g per kilogram of flour), and the amount of sugar must be raised to 11% (110 g per kilogram of flour). Even under identical temperature conditions, it is also necessary to plan for a slightly longer second fermentation—an increase from 80 minutes to 90 minutes—or to allow for 75 minutes if the croissants are placed in a proof box set at 33°C (91°F).

It is possible to vary the basic procedures in several different ways. For example, from the same formula batch—either butter or margarine—the baker may modify the first fermentation by having part of the dough rise under ambient temperature while the remainder is placed under refrigeration for longer or shorter periods of time. After lamination and makeup, croissants may also be frozen for a maximum of 4 to 5 days before being thawed in the refrigerator or at ambient temperature, placed in a proof box to rise, and then finish baked at the scheduled time. When properly used, this system can permit a great degree of flexibility while producing croissants of excellent quality.

Croissants may also be produced by the use of a temperature-controlled proofing cabinet. The dough for this application generally has a slightly higher level of both yeast and ascorbic acid. The level of the latter ingredient may increase from 20 to 25 or even 30 ppm.

The formula in Exhibit 13-3 allows good and quick production of croissants that are highly distinctive and that achieve a rare quality level. Croissants made from this dough, which has a high percentage of organic acids, have the advantages of both excellent flavor and very good shelf life. In addition, this formula is very well suited to the production of fully formed croissants either for the *pousse controlée* method with a temperature- and humidity-controlled proof cabinet, or for freezing during a 4- to 5-day storage period, as is described below.

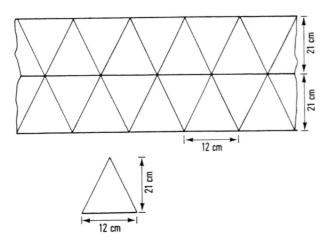

Figure 13-4 Croissant Cutting Diagram.

Figure 13-5 Cutting Croissants by Hand and with a Cutter.

Production of Croissants Entirely Under Refrigeration

Today many professional bakers produce croissants completely under refrigeration. This results in very attractive croissants, but unfortunately they exhibit some serious failings. Since the first part of the alcoholic fermentation is suppressed, the resulting dough has a low organic acid content. This fault seriously affects the taste and keeping qualities of the product, especially when the croissants are made according to the straight dough method.

The dough is generally mixed the day before use, to a final temperature of between 18 and 20°C (64.4°F) from the mixer. It is divided into dough blocks from 2

Yeast-Raised Sweet Doughs 147

Exhibit 13-3 Croissant Formula with the Addition of Prefermented Dough

Formula

	Weight		Fermented Dough	Final Dough
Flour	2.0 kg (70.55 oz)		600 g (21.16 oz)	1.4 kg (49.38 oz)
Water	1.2 l (42.33 oz)	60%	415 g (14.64 oz)	785 g (27.69 oz)
Salt	48 g (1.69 oz)	2.4%	12 g (0.42 oz)	36 g (1.27 oz)
Yeast	70 g (2.47 oz)	3.5%	10 g (0.35 oz)	60 g (2.12 oz)
Sugar	180 g (6.35 oz)	9%		180 g (6.35 oz)
Milk powder	30 g (1.06 oz)	1.5%		30 g (1.06 oz)
Malt extract	25 g (0.88 oz)	1.25%		25 g (0.88 oz)
Butter	40 g (1.41 oz)	2%		40 g (1.41 oz)
Ascorbic acid	(30 mg)	(15 ppm)		(30 mg)
FERMENTED DOUGH			1037 g (36.58 oz)	(1037 g [36.58 oz])
TOTAL DOUGH WEIGHT				3.593 kg (126.74 oz)

Procedures

	Fermented Dough	Final Dough
Mixing: Low speed	3 min	4 min
Mixing: Second speed	8 min	5 min
Dough temperature from mixer	(24° C/ 75° F)	(20° C/ 68 ° F)
Rest period (fermented dough)	4 h @ 24° C or 75° F *or* 15 h @ +4° C / 39° F	
Cold first fermentation @ +2° C/35.6° F		75 min
Lamination		35 min
Dough cutting & makeup		25 min
Second fermentation @ 27° C/80.6° F		75 min
Baking @ 230° C / 446° F		20 min
PRODUCTION TIME		4 h

---

to 3 kg in weight before being placed under refrigeration at 1 to 2°C (35.6°F), for 15 to 18 hours. Following refrigeration, the dough pieces may be laminated into separate batches of croissants over several hours until they have all been used.

The formed croissants are placed on baking sheets, covered with heat-resistant paper, and stored at 0°C (32°F) from 12 to 48 hours in low-temperature proofing cabinets. Normal procedures are used for final proofing and baking.

The Production of Unbaked Frozen Dough Croissants

Freezing has been used in the production of croissants for several years. (Freezing may be used at several different stages in the overall process.) It may, of course, occur after the croissants are fully formed: these are raw (unbaked) frozen dough croissants, which must be thawed and finish proofed before baking. Freezing may also take place at the end of the second fermentation period: these are fully proofed raw (unbaked) frozen dough croissants, which may be fully baked (or partially baked) directly upon being removed from the freezer.

Production of frozen unbaked dough croissants may take place over a period of 4 to 5 days maximum, as noted previously. Nearly any type of croissant formula may be used successfully in this instance.

However, the desired frozen storage period may be increased to several weeks or even to several months. These lengthened storage periods require the use of specially adapted formulas. In comparison to conventional production,

- the flour used should be slightly stronger, and its enzymatic activity should be at the low end of the normal range;
- the yeast level should be around 70% higher;
- the amount of ascorbic acid should be doubled, or sometimes tripled;
- a small amount of fat should still be added during mixing;
- neither milk powder nor malt extract should be used;
- dough temperature at the end of mixing should be equal to or less than 20°C (68°F);

Dough lamination should be preceded by a 10-minute rest period under refrigeration. The lamination stage should be composed of three "pocketbook" folds and two double "book" folds (see illustrations).

Shaping of the product should take place immediately following lamination, and the formed croissants should be placed in a freezer enclosure maintained at –40°C until the dough reaches an interior temperature of –18°C to –20°C (–2° to –4°F). At that point the croissants should be placed in plastic bags to protect them from the drying effects of air and then stored at –15°C to –18°C (–4° to –1°F). They may be maintained in the frozen state as required for up to several weeks before shipping and distribution. They are then thawed, proofed, and finish baked in retail croissant shops, restaurants, and sometimes even bakeries.

Since the first fermentation is reduced as much as possible before freezing, frozen croissants only undergo what is equivalent to the second fermentation in the proof box. The resulting product is generally satisfactory as far as appearance alone is concerned. However, these croissants must be consumed while very fresh, and even then they are inferior in taste to croissants made by conventional methods. They are also characterized by an inferior shelf life.

The production of croissants from raw, fully proofed dough allows storage of frozen product for periods extending over several weeks as well. Although such croissants are more difficult to store (because of their greater volume and fragility), they offer a great advantage in that they may be baked on demand.

The formula for this product may also be improved, but to a lesser extent than the preceding, by the use of a good flour and a higher than normal level of both yeast and ascorbic acid. Furthermore, it will accept the addition of 15% to 20% of prefermented dough, added during the mixing stage. If the straight dough method is used instead, the first fermentation should be allowed to proceed only for a short time. In both cases, the dough should be rapidly and adequately cooled before lamination. The proof level reached by the formed croissants before freezing should also be systematically adjusted so that it is slightly less than the level used in traditional production.

Furthermore, whether the croissants are placed in the oven immediately upon being removed from freezer storage (at the previously mentioned –15°C / –4°F), or after having been thawed for a brief time at room temperature, they should be baked at a rather low temperature [230°F] to ensure a gradual thawing and to achieve strong oven spring, good development, and good volume. For these reasons, the croissants should remain in the oven for a relatively long time—about 25 minutes.

The quality and shelf life of croissants made according to these procedures are generally very satisfactory, thanks to the beneficial effects of a small amount of prefermented dough or the favorable results of the straight dough first fermentation.

CHOCOLATE-FILLED BUNS FROM CROISSANT DOUGH

These small buns (called *petits pains au chocolat*) are usually made from croissant dough, but they may also be made from milk roll dough or brioche-style bread dough. With croissant dough, the dough is rolled out into a rectangle 3 mm thick, while with milk roll dough the dough rectangle should be 5 mm thick.

The next step is to cut the dough rectangle into bands that are the same width as the small chocolate bar that is to be used as the filling. The bar is placed flat on the dough band, which is then rolled around it so as to give two complete thicknesses of dough around the bar. Makeup of this product is then completed by cutting the dough roll with a knife into individual units. The roll units are placed on baking sheets, given a medium proof, brushed with a whole

---

Croissants rise better in the oven when the point of the triangle is on top of the croissant, rather than underneath. In Paris (but not elsewhere in France) croissants made with butter are left straight, rather than being bent into a crescent shape (Figure 13–6). This differentiates them from croissants made with margarine.

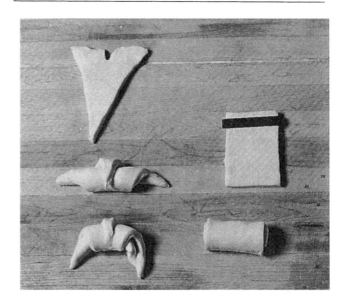

Figure 13–6 Croissant and Pain au Chocolat Production.

egg wash, and baked at the same temperature (230°C) as used for croissants (Figure 13-7).

SNAIL ROLLS

This product (also called *schneckes* or *schnecken*) may also be made either from croissant dough or from milk roll dough. In either case, a dough piece weighing from 1 to 1.5 kg is rolled out to a rectangular shape up to 6 mm thick. The dough rectangle is then cut into bands between 20 and 25 mm wide. After cutting, the dough strips are brushed with melted butter or pastry cream, sprinkled with granulated sugar, and garnished with sorted, cleaned Corinth raisins.

The dough band is then rolled up, and it should have a diameter of about 7 cm. The dough roll is sliced into individual units about 15 mm in length. These are then placed on lightly greased baking sheets so that one of the flat (cut) sides is uppermost, given a medium proof, and brushed with beaten whole egg wash (Figure 13-8). They may be sprinkled with granulated white sugar before being placed in a rather cool oven to bake (around 210°C / 410°F).

BRIOCHES

Conventional Paris-Style Brioches

This typically French product is one of the oldest examples of yeast-raised sweet goods. Its origins are lost so far back in the depths of time that it would be foolish to try to determine exactly the period during which it was developed. It is made from an extremely

---
■

The brioche is a wonderful baking specialty that does not have the following it deserves. Perhaps that is due in part to a simple misunderstanding. Marie Antoinette did not say "Let them eat cake," but instead, "Let them eat brioche." The rest is history.

Figure 13-7 (A) Baked Croissants and (B) Pains au Chocolat.

■

Appropriate size chocolate pieces are commercially available. High quality bitter chocolate is a wise investment. When making *pains au chocolat*, some Japanese bakers spoil their customers with two pieces of chocolate.

Figure 13-8 Schneckes and Brioche Royale.

rich dough, with a high proportion of butter, and the flour is traditionally moistened only with eggs.

The aroma and flavor of the brioche come primarily from the dual presence of eggs and, especially, butter. This is in spite of the fact that in former times it may have been flavored with orange flower water. (A painting completed during the 18th century reign of Louis XV by the painter Chardin shows a brioche into which the stem of an orange flower has been inserted.)

It is highly likely that the brioche of that time contained a smaller proportion of butter or was even made with animal fat. It may also have been sweetened with honey, and the aroma of the orange flower water was important to the enhancement of the honey flavor. It should also be noted that regional brioches still continue to be flavored with orange flower water to some extent (see Chapter 14).

The overall taste of the brioche—including aroma, flavor, and mouthfeel—is very strongly influenced by the quality of the raw materials from which it is made. It is thus very important to choose them with extreme care.

With a few rare exceptions, the flour used for brioche production has always been white wheat flour: in the days of grist milling this was the finest quality "flower of the wheat," and later high-quality flour ground from wheat farina was used. The flour used today should be a pure wheat flour (without added bean flour or malt). It should have a W value equal to or greater than 220. The gluten should be firm and extensible, and present at a rate equal to or greater than 10.5%.

In ancient practice, leavening was always by means of a natural flour ferment or "sourdough." When baker's yeast became available, it became the preferred leavening method. However, the use of flour ferments has not completely disappeared, and some professional bakers continue to use a rapid variation of this ancient method as the sponge and dough. Today the straight dough method is the most commonly used, although the sponge and dough is still one of the better methods.

The butter used should be well pressed or kneaded, with a good aroma. If margarine is used instead, it should be a vegetable oil–based margarine with a fusion point between 32 and 36°C (89.6°and 102.2°F). The use of vegetable margarine allows the use of a slightly weaker flour, while still yielding brioches that have excellent volume. However, the delicacy and flavor will be inferior to those of authentic butter brioches. It is possible to partially remedy the taste deficiency of margarine brioches by slightly increasing the sugar content. The eggs used in brioche production should be relatively fresh and in good condition. Milk is sometimes added to temper the flavor of the egg yolks, which is sometimes too strong when eggs are used alone, and the use of a little milk—generally around 15%—is not undesirable, since milk creates a more agreeable and better balanced flavor profile. If raw milk is not available, the baker should use fresh, pasteurized liquid milk.[4]

Formula, Procedures, and Leavening of Brioche Doughs

Brioche doughs may be leavened by the straight dough method (Exhibit 13–4), by the addition of prefermented dough (Exhibit 13–5), or even by use of the sponge and dough method (Exhibit 13–6). However, if all other factors are equal, leavening by either the prefermented dough method or the sponge and dough procedure will ensure more consistent quality and more rapid production.

To use salted butter with this and other brioche recipes, reduce the salt from 2.4% (120 g; 4.23 oz) to 2% (100 g; 3.53 oz)

Exhibit 13–4 Formula and Procedure for Brioche by the Straight Dough Method

Formula

Flour	5 kg (176.37 oz)	100%
Salt	120 g (4.23 oz)	2.4%
Yeast	200 g (7.05 oz)	4%
Sugar	600 g (21.16 oz)	12%
Butter	2.25 kg (79.37 oz)	45%
Eggs	2.5 kg (88.18 oz)	50%
Milk	400 g (14.11 oz)	8%
Ascorbic acid	100 mg	20 ppm
	(40 mg)	(20 ppm)
TOTAL DOUGH	3.488 kg	

Procedures

Mixing: 1st speed	3 min
Beater: 2nd speed	12 min
Dough temperature	(22° C / 71.6° F)
1st fermentation	3 h
Punch down after	1 h 30 min
2nd punch after	2 h 15 min
Divide, form	40 min
2nd fermentation	80 min @ 27° C / 80.6° F
Baking (220° C / 428° F)	15 to 30 min
TOTAL TIME	5 h 45 min

Exhibit 13–5 Brioche Formula with the Addition of Prefermented Brioche Dough

Formula

	Weight		Fermented Dough	Final Dough
Flour	5.0 kg (176.37 oz)		1.0 kg (35.27 oz)	4.0 kg (141.09 oz)
Salt	120 g (4.23 oz)	2.4%	24 g (0.84 oz)	96 g
Yeast	200 g (7.05 oz)	4%	40 g (1.41 oz)	160 g (5.64 oz)
Sugar	650 g	13%	130 g (4.59 oz)	520 g (18.34 oz)
Butter	2.25 kg (79.37 oz)	45%	450 g (15.87 oz)	1.8 kg (52.91 oz)
Eggs	2.25 kg (79.37 oz)	45%	450 g (15.87 oz)	1.8 kg (52.91 oz)
Milk	750 g (26.46 oz)	15%	150 g (5.29 oz)	600 g (21.16 oz)
Ascorbic acid	(75 mg)	(15 ppm)	(15 mg)	(60 mg)
FERMENTED DOUGH			2.244 kg	(2.244 kg)
TOTAL DOUGH WEIGHT				11.22 kg

Procedures

Mixing: Low speed		3 min	3 min
Mixing: Second speed		12 min	12 min
Dough temperature from mixer		(22° C / 71.6° F)	(22° C / 71.6° F)
First fermentation @ +2° C / 35.6° F			100 min
Punch down (degassing) after			50 min
Rest period for fermented brioche dough		5 h @ 24° C / 75.2° F or 15 h @ +4° C / 39.2° F	
Dough division & makeup*			40 min
Second fermentation @ 27° C / 80.6° F			80 min
Baking @ 220° C / 428° F			15–30 min
Production time:			4 h 25 min

*Brioche dough is most often refrigerated at the end of the first fermentation. When it is to be used, an appropriate amount of refrigerated dough is placed on a work table and beaten with a rolling pin to restore elasticity, then weighed and shaped.
The second fermentation is at the normal 27° C, but is a bit longer than is usual for other breads because the dough is cold.

Exhibit 13–6 Brioche Formula Leavened by the Sponge and Dough Method

Formula

	Weight		Yeast Sponge	Final Dough
Flour	5.0 kg (176.37 oz)		1.25 kg (44.09 oz)	3.75 kg (132.28 oz)
Salt	120 g (4.23 oz)	2.4%		120 g (4.23 oz)
Yeast	175 g (6.17 oz)	3.5%	175 g (6.17 oz)	
Sugar	600 g (21.16 oz)	12%		600 g (21.16 oz)
Butter	2.25 kg (79.37 oz)	45%		2.25 kg (79.37 oz)
Eggs	2.25 kg (79.37 oz)	45%		2.25 kg (79.37 oz)
Milk	750 g (26.46 oz)	15%	750 g (26.46 oz)	
Ascorbic acid	(75 mg)	(15 ppm)		(75 mg)*
YEAST SPONGE			2.175 kg (76.72 oz)	(2.175 kg [76.72 oz])
TOTAL DOUGH WEIGHT				11.145 kg (393.12 oz)

Procedures

	Yeast Sponge	Final Dough
Mixing: Low speed	6 min	3 min
Autolysis rest period		25 min
Mixing: Second speed		10 min
Dough temperature from mixer	(24° C / 75.2° F)	(22° C /71.6° F)
Rest period for yeast sponge	45 min	100 min
First fermentation @ +2° C / 35.6° F		50 min
Punch down (degassing) after		40 min
Dough division & makeup		40 min
Second fermentation @ 27° C / 80.6° F		80 min
Baking @ 220° C / 428° F		15–30 min
PRODUCTION TIME		4 h 48 min

* The original formula stated 750 mg, which is obviously an error.

Brioche Leavened by Addition of Prefermented Dough

In this instance, the prefermented dough added during mixing is a remainder of brioche dough from a previous batch, which usually has been kept under refrigeration. The addition of this prefermented culture allows the baker to reduce the length of time required for dough maturation before forming it into brioches. This represents a valuable saving in production time without any negative effect on quality. However, since this long fermentation involves the exhaustion of the sugars added to the original batch during mixing, it also requires a slight increase in the quantity of sugar added to the final dough.[5]

Makeup and Correct Use of Brioche Dough

Brioche dough is rather soft in consistency, but requires a thorough mixing. The formation of a fully developed gluten network is rather laborious and time-consuming because of the use of eggs and a small amount of milk as a wetting agent, the addition of a high percentage of sugar, and the generally weak structure of the mixture. For these reasons, it is advisable not to include all the eggs at the beginning of mixing. About 1/8 of the total should be held back and should be added when the dough has developed sufficient structure.

Furthermore, whenever prefermented brioche dough or a yeast sponge is to be added, this should be done halfway through the mixing stage. Mixing should then continue until the dough becomes homogeneous and begins to detach itself from the sides of the mixing bowl. It is only at that point that the butter, preferably kneaded in advance, should be incorporated into the dough. The dough will absorb the butter very rapidly, and mixing should again continue until the dough becomes very smooth, begins to develop body or structure, and again detaches itself from the sides of the mixing bowl.

The autolysis rest period is at least as important for brioche dough as it is for bread and croissant dough—unless it is greater. It allows a reduction in the duration of mixing and at the same time reduces the oxidation, bleaching, and debasing of the dough. This results in better preservation of the aroma and flavor provided by eggs and butter.

Mixing should be carried out at first (low) speed in several stages: in the first, flour, sugar, eggs, and milk are blended together. In the second stage, the salt, yeast, and ascorbic acid are added and mixed into the forming dough. In the third stage, the yeast sponge is incorporated into the dough. Butter is added separately as the final ingredient.

The baker must take care to ensure that the dough has reached the necessary stage of development and maturity before division and dough forming, since a certain degree of dough strength is necessary in these operations. As noted earlier, degassing or "punching down" the dough during the first fermentation improves dough cohesion and helps give it the necessary strength before forming or molding. The dough should be degassed twice when the straight dough method is used, and at least once when a prefermented culture is used.

Refrigerated storage or retarding is commonly used in the production of brioches. The most common practice is to retard the dough during the first fermentation. After allowing the dough to rest for an hour at ambient temperature, the dough is placed in refrigerated storage at temperatures from 2 to 4°C (35.6°F to 39.2°F). It may then be retarded for 6 to 18 hours or more.

When the dough is removed from the retarding chamber it initially has a dull, lifeless appearance. The condition of the dough is improved by rolling it out and lightly laminating it in order to form (or reform) the gluten network. It is then given a 30-minute rest period at ambient temperature (from 22 to 24°C / 71.6°F to 75.2°F), or for 45 minutes under refrigeration (from 10 to 15°C / 50°F to 59°F). The dough, which is rather elastic, is then divided into pieces of an appropriate size, rounded, molded, and then placed in molds or on baking sheets.[6] Following these procedures ensures that dough forming will be easier and that the appearance and shape of the molded dough pieces will be improved. The end result is an increase in the volume of the brioches after baking and improvement of the lightness, smoothness, and softness of the crumb.

Brioches may also be placed in temperature-controlled proofing cabinets after forming, which allows them to be baked 12 to 48 hours later. In this procedure they are placed in cabinets cooled to 0°C. At the end of the storage period, the temperature is raised to 27 to 28°C (80.6°F to 82.4°F) for 2 to 2 1/2 hours, after which they are ready to bake.

Finally, brioche dough may be frozen either before or after forming:

- *Before forming:* After having fermented from 1 hour 30 minutes to 2 hours, the dough is divided into large pieces of around 4 kg in weight. These are rolled out, placed on baking sheets, covered with protective sulfur-treated paper, and frozen

at −20°C (−4°F). They may be stored in this state from 7 to 10 days. They may be thawed at ambient temperature as required, then divided, formed, given an appropriate proof, and baked.
- *After forming:* the brioches may be frozen in a freezing chamber at −40°C (−40°F), then stored at −15°C (−4.3°F) for 10 days to several weeks. They are thawed on demand for about 3 hours at ambient temperature, proofed at 27 to 28°C (80.6°F to 82.4°F) for around 90 minutes, and then baked normally.

When the storage period may last more than several weeks, modifications to both the formula and procedures are required: more yeast and ascorbic acid are needed, the use of the straight dough method is mandatory, and the first fermentation period must be almost completely eliminated. Furthermore, it becomes more difficult to produce a quality butter brioche, and it is sometimes preferable to use margarine instead.

The Different Brioche Types

Several different types of products may be produced from common brioche dough (Figure 13–9).

Figure 13–10 Shaping Brioche.

Well-made brioches are an excellent product and the pride of many French bakers. They merit more attention in North America.

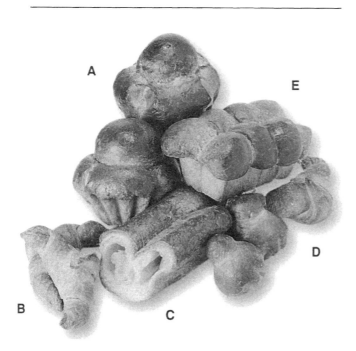

Figure 13–9 Brioche Shapes. (A) Large Brioche Parisienne, (B) Croissant brioché, (C) Brioche feuilletée, (D) Small brioché à tête, (E) Brioche de Nanterre.

Different makeup and forming methods are used for each of these brioche products, and their individual names are derived from each product's distinctive appearance and physical characteristics (Figure 13–10). In addition, both the butter content and consistency of the dough may vary considerably.

Some of the more widely known brioche variations are made from the same basic dough: small and large brioches with "heads" or topknots; the Nanterre-style brioche; the crown brioche; the mousseline or chiffon brioche, which has a very high butter content; the royal brioche, to which candied fruit has been added; the laminated dough brioche, which combines common brioche dough with a high proportion of laminating fat; the brioche croissant, which has a higher than normal proportion of butter in the dough combined with a more normal level of laminating fat; and a lower-fat, lower sugar dough made expressly for the Lyon-type sausage brioche. Formulas that have been slightly adapted to the production of the sausage brioche is given in the text.

Brioche with a "Head" or Topknot

This is the most widely known version of the brioche and is recognized as the "traditional" form. The formula for this product is of "average" richness, since it has a butter content that ranges from 40% to 50% based on flour weight. These brioches must be made from a rather strong dough.

The raw dough is portioned and weighed at between 35 and 40 g (1.23 and 1.4 oz) for the small version, and between 200 and 300 g (7 to 10.6 oz) for the larger brioches. Both are generally baked in round-bottomed, flared molds with fluted sides. These brioches are made up by dividing the weighed dough into two portions: the "body" and the smaller "head." After baking, the latter forms a rounded head—either slightly inclined or upright—surmounted by a slightly inflated "cap." The brioches are brushed with whole egg wash before baking and are loaded into the oven on preheated baking sheets.[7]

The Crown Brioche

This type of brioche is generally made from the same dough as the preceding but uses a dough weight that is equal or greater than 250 g. After the dough has been divided and weighed, each section is rounded into a ball. The dough ball is allowed to rest for 15 minutes, then pierced along the axis. From that point the hole is progressively enlarged until the desired "crown" form is achieved. The formed dough "crown" is then placed on a baking tray and then into a proofing chamber, where it is protected from the air. Before being placed in the oven, the crowns are brushed with an egg wash, following which the tops of the loaves are cut into "saw teeth" with scissors. They may also be sprinkled with sugar after baking.

The Nanterre-Type Brioche

This type of brioche should be made from a soft dough that has a relatively high butter content, such as the "high average" level of 50% on flour weight noted for the *brioche à tête*. The dough should also be fairly strong. This brioche is baked in rectangular cake pans.[8] The 250 to 300 g dough piece, which is the weight of the whole brioche, is further subdivided into 6 or 12 identical smaller pieces. These are rounded into balls and placed either side by side or in alternating positions to form two rows in the cake pan.

An alternative form is composed of only a single row of dough balls. The top of each is cross-cut with scissors before baking. The dough should be placed into the pan so that only about one fourth of the volume of the pan is used. The dough balls should be brushed with egg wash before baking in this case as well.

The Mousseline or Chiffon Brioche

This is the richest brioche and is made with at least 700 g butter per kilogram of flour. For this reason a superior quality flour is absolutely required. In addition, the use of refrigeration is practically indispensable during the summer, in order to obtain a product with good structure and appearance from a dough that contains so much butter. Good dough strength and maturity are also necessary in any season.

In production, the dough is rounded and placed in tall, smooth-sided cylindrical molds, with the molding or rounding seam at the bottom of the mold. At the end of the proofing period, the dough should reach the upper rim of the mold. The resulting dome is brushed with egg wash, then a cross shape is cut into the dough with a pair of scissors.

Before the brioche is placed in the oven, the mold is generally wrapped with a sheet of treated paper to form a collar that extends several centimeters beyond the mold. This collar ensures that the oven spring development of the brioche will be vertical. Baking takes place at a very moderate temperature, around 190°C.

The Royal Brioche

This variation is made from a good-quality, strong brioche dough, which must tolerate dough forming and the filling of candied fruit (see Figure 13–8). Forming begins with rolling the dough into rectangles about 15 cm wide and about 12 to 15 mm thick. The baker brushes the dough sheet with a layer of pastry cream and then arranges candied fruit on the pastry cream. The dough sheet is then rolled into a "sausage" or cylinder, and the dough roll is then sliced into portions about 2 cm thick. The rolled sections are then gently deposited into a previously greased Genoa (sheet) cake pan with the edges of the slices touching one another.[9]

After reaching an appropriate level of proof, the brioche is brushed with whole egg wash and baked at a moderate temperature without steam. After baking, royal brioches are brushed with sugar syrup.

The Laminated Brioche

As might be imagined from its name, the flaky brioche is made from a common brioche dough that undergoes lamination with additional butter after the first fermentation. It can also be made with a specially made dough that has a lower butter content.

The dough is cooled to 10 to 12°C / 50°F to 53.6°F) then rolled out lightly. The desired amount of cold kneaded butter is enfolded into the dough sheet, and lamination proceeds according to the "classical" method. As in the case of croissants, the baker may choose to give three pocketbook folds or two book folds.

The amount of butter used in lamination is about 300 g per kilogram of dough, and this use level generally gives very, very good results. After a second proof period of about 2 1/2 hours, the dough may be baked "as is" without further forming and yields a very fine product of exceptional quality. This dough is suitable for making Danish pastries, filled with either creams or jams. When the fillings are comparable in quality to the dough, the resulting pastries are truly superior.

The Brioche-Style Croissant

The brioche-style croissant is particularly rich, considering both the amount of fat that is mixed into the dough and the additional fat which is used for lamination. At the present time this product is made only rarely.

A typical formula for a brioche-style croissant might include the following ingredients for each kilogram (35.27 oz) of strong flour: 22 g (0.78 oz) salt, 30 to 40 g (1.25 oz) fresh yeast, 90 g (3.17 oz) sugar, 250 g (8.82 oz) butter, 8 eggs, and 150 g (5.29 oz) milk. The dough should be very slightly firm at the end of mixing.

The first fermentation should be around 2 hours and should include a single degassing or "punching down" of the dough. At the end of the first fermentation, it is preferable to cool the dough at a temperature of from 16 to 18°C (60.8°F to 64.4°F).

At the end of this refrigerated rest or maturation period, the dough is rolled into a rectangle and laminated with one single fold and one book fold, using 250 g (8.82 oz) butter for each kilogram (35.27 oz) of dough. After another brief refrigerated rest period, the dough is rolled to a rectangle about 3mm thick, cut into triangles, rolled into croissants, and placed on a baking sheet.

After rising for 60 to 70 minutes in a proof chamber, the croissants are brushed with egg wash and baked at 210°C / 410°F). An icing is generally applied to these special brioche croissants before sale.

The Lyon-Style Sausage in Brioche

The brioche dough used to contain cooked Lyon sausage is notably less rich in sugar and slightly lower in butter than common brioche dough (Exhibit 13–7). For 1 kg (35.27 oz) flour, reduce sugar to 60 g (2.12 oz), butter to 400 g (14.11 oz), and eggs to 350 g (12.35 oz). In addition, 200 g (7.05 oz) milk are used in the dough. The dough will be slightly more firm than that of a conventional brioche and should be relatively strong.

Preparation. Marinate the meat and the fat together 24 to 48 hours with the ingredients noted above, except for the truffles, pistachios, and starch (if used). Pass the salted mixture through an 8mm grinder to finish this preparatory step. Then mix with the binding agent (starch if used), then the truffles and pistachios. Stuff the mixture into a pork casing or into the straight "Naturin®" (an artificial type) casing: allow to dry naturally or in a drying chamber.

Cooking. Heat the cooking liquid to boiling and set temperature at 90°C / 194°F. Drop in the sausages, then set temperature at 85°C / 181.4°F. Allow the temperature to drop to 75°C / 167°F, then cook for 40 minutes for sausages 35 to 40 mm in diameter.

Preparing for the Baking Mold. After cooling, remove the casing from the sausage. Coat the sausage completely with whole egg wash and coat the entire surface with flour. Place the coated sausage on a previously prepared rectangle of brioche dough. Roll the dough around the sausage, pinch shut the seam and the ends of the dough. Insert support hooks into the ends of the sausage as illustrated.

Place the wrapped sausage into the previously greased mold, with the tightly shut seam down and the supporting hooks resting on the ends of the baking pan. Allow the dough to rise for 60 to 70 minutes, then brush the top of the dough with egg wash and bake the sausage-filled brioche for 30 to 35 minutes at 220°C / 428°F.

Exhibit 13–7 Formula and Procedure for Cooked Lyon Sausage*

Quality lean pork (boneless pork shoulder, well trimmed of any fat)	1.5 kg (52.91 oz)
Fresh, firm-textured back fat (fresh bacon)	500 g (17.64 oz)
Total meat	2 kg (70.55 oz)
Curing salt—(a mixture of salt and sodium nitrite available from sausage supply houses and larger supermarkets)	20 g (0.71 oz)
Common table salt (Sodium chloride)	20 g (0.71 oz)
White pepper	4 g (0.14 oz)
Saltpeter‡	1 g (0.035 oz)
Sugar	8 g (0.28 oz)
Nutmeg	1 g (0.035 oz)
Green pistachios (unsalted, peeled)	40 g (1.41 oz)
Black truffles, chopped	60 g (2.12 oz)
Cognac	60 g (2.12 oz)
Cornstarch (optional)†	40 g (1.41 oz)
2" diameter beef casing	As needed

*In the U.S., it is generally forbidden by law to produce meat products in the same wholesale facility as bakery products, due to the danger of cross-contamination of finished product. Wholesale bakers who wish to produce this specialty brioche should contract with a certified meat processor to produce the cooked sausage. Many restaurants and retail bakery establishments are exempt from this regulation. Check local food safety codes as applicable.

†Starch is only rarely encountered in small scale sausage production.

‡Most recipes include either curing salt or saltpeter. U.S. regulations usually require curing salt.

The Lyon sausage recipe that Professor Calvel provided in the original French text is a capsule summary. For most readers it would not provide enough detail to successfully produce the sausage used in the Lyon-style "sausage in brioche." However, a direct translation from the French text is given in order to maintain integrity with the original. Although kielbasa is sometimes used as a substitute in this application, it is not as delicate as the sausage formulation. The following provide more detailed procedures:

Initial preparation. The lean pork and back fat must be of impeccable quality and freshness. Premix the salt, curing salt, saltpeter, and sugar. Cut the lean pork and back fat into 1-inch cubes. Place the meat and back fat cubes in a mixing bowl with the premixed ingredients. Stir the meat cubes and premix together until the pieces are well coated. Refrigerate the coated meat cubes at 4°C / 39°F for 24 to 48 hours.

Grinding. **It is extremely important to keep the meat mixture very cold during the grinding, mixing, and stuffing procedures** in order to maximize quality and minimize the chances of spoilage and food-related illness.[10] Grind the pork / pork fat mixture in an effective and well-maintained meat grinder equipped with an 8 mm grinding plate. Refrigerate the meat mixture as necessary to maintain the 4°C / 39°F temperature.

Mixing. Add the pepper, nutmeg, pistachios, truffles, and cognac to the ground meat and knead the mixture until homogeneous. Refrigerate as necessary to maintain 4°C / 39°F temperature.

Casing preparation. Rinse the interior and exterior of the beef casings with cold water, then soak them in cold water for 10 to 15 minutes. Drain the casings, then cut them into 30 cm / 12 inch lengths. Tie off one end with butcher's twine.

Stuffing. Place the ground meat mixture in the feed hopper of a sausage stuffer equipped with a spout of appropriate diameter. Stuff the sausage mixture into the casings, tie off the end of the sausage with twine, leaving enough twine to make a loop from which the sausage may be hung.

Curing. Hang the sausages in a warm (25° to 30°C / 77° to 86°F), draft-free, and fairly humid environment until they become reddish-pink and the casings are dry. Refrigerate, then use within 7 days.

NOTES

1. In much of Europe the mixing and baking areas were traditionally located below ground, just as they were in the United States until the early part of the 20th century.
2. Croissants and other free-standing pieces are baked on sheet pans lined with nonstick silicone-coated paper.
3. Egg wash is normally made from whole egg yolks beaten with salt, or less frequently, egg yolks, water, and salt.
4. Since this book was written European food laws have changed. The direct use of raw milk in food products is almost universally prohibited by law—including such use in North America—in both wholesale and retail baking as well as many other types of food production. The recently appearing danger of serious food poisoning at the consumer level from contamination of finished foods by *Salmonella, Listeria, E. coli,* and a number of other potentially deadly organisms is simply too serious to ignore. Improperly handled milk products, contaminated ground meats, and unsanitary shell eggs are the most common causes of this problem, and all can have serious implications for both the craft and home baker.
5. It is not Professor Calvel's intention that bakers prepare a batch of brioche dough which is to be used as the prefermented dough especially for this recipe. This is a method for either using up some extra dough, or using an insufficient amount of dough on hand as a base for a greater amount of dough required for an unexpected order. If a preferment is to be made expressly for a brioche recipe, the sponge and dough recipe

(Exhibit 13–6) is much simpler choice yielding equal results. Also note that the use of brioche dough as a perferment may be done occasionally but not systematically and perpetually. This would lead to off flavors.

6. The molds or tins used for brioches are brushed with melted butter before the dough is placed into them.

7. For an extra-glossy finish, large brioches and other fancy pieces may be coated once with egg wash 15 minutes before baking, and a second time immediately prior to baking.

8. Pans such as those commonly used for pound cake.

9. Much as with American cinnamon rolls.

10. Lower grinding and mixing temperatures help to produce the desired firm texture in the finished sausage.

Regional Brioches

CHAPTER 14

- Regional brioches.
- Vendée-style brioches.
- Specialty brioches.
- Brioche-type hearth cakes from other lands.

REGIONAL BRIOCHES

Although it was the ancestor of the Parisian brioche that inspired the painter Chardin in the 18th century, it is not—and was not—the only type made in France. Other regional brioches may be made during certain seasons of the year—or even all year long—and are certainly worthy of being better known. That is especially true of the Vendée-type brioche, the *gâche*, certain types of Three Kings' cakes, the Palm Sunday and Rodez hearth cakes, the "fist" from Romans-sur-Isère, and the kugelhopf. Similar products are also made in other countries, such as the Spanish *mouna* (which is also made in France), the Italian *panettone*, and Argentine *pan dulce* or sweet bread.

In the past, most of these brioches were produced from a naturally fermented culture (sourdough). A few are still made in this traditional way even today, and the distinctive character of such products is much appreciated by consumers. However, except in the case of the Italian *panettone*, this method is in decline, and even in those instances where it continues to be used, it is often combined with fermentation by baker's yeast. This tendency explains why the more practical sponge and dough method is replacing the natural fermentation method in many cases.

In order to avoid making the text too dense and difficult to read, I will supply a single example of the use of a naturally fermented culture that is applicable to several different brioches, such as the hearth cakes of Rameaux and Rodez, and the pogne of Romans-sur-Isère; the composition and makeup procedures are very similar. However, I will give individual formulations and procedures for those made by the sponge and dough and straight dough methods, since each brioche has its own specific leavening techniques.[1]

NOTE: The mixing times in this chapter are for planetary mixers.

Kugelhopf

The kugelhopf[2] is an Alsatian cake or hearth cake very similar to the traditional brioche. This traditional yeast-raised, sweet dough cake is made in Alsace by housewives, bakers, and pastry chefs and may be encountered at all types of celebrations, including family gatherings, village holidays, and receptions. It always goes well with the excellent white wines of Alsace.

The dough is quite rich and contains raisins, but part of the eggs are replaced by milk (Exhibit 14–1 and Exhibit 14–2). The formed loaves are baked in partially glazed clay molds lined with almonds or other nuts. These molds give to the kugelhopf a particular aroma, which is every bit as seductive as it is distinctive, from the fat that is absorbed by the clay of the unglazed portion of the mold. After oxidation the absorbed fat imparts a slightly rancid scent to the mold, and this is discreetly passed on to the dough during baking. This scent manifests itself as a light perfume that has no real comparison, greatly improving the aroma and the flavor of the kugelhopf and giving it a highly original and especially agreeable quality.

The raisins should be soaked in schnapps (a liquor made from a mixture of grapes and fruit) or, if that is not available, in **marc** or **grappa,** a brandy made from spent, fermented grapes or skin pressings from the winemaking process. The almond halves or nut kernels are placed in the bottom of the molds, and after the cakes are baked and unmolded they will be found encrusted into the upper surface of the kugelhopf (Figure 14–1).

After unmolding and while the cakes are still warm, they may be improved by being brushed with melted butter. This has the dual effect of increasing the aroma of butter absorbed into the dough from the mold, and also helps to seal the exterior surface of the cake and thus to improve its keeping qualities. It further helps the adhesion of the icing sugar with which the kugelhopf is always covered before being served. Finally, in order to really appreciate good kugelhopfs, it is best to eat them 24 to 48 hours after baking.

Bacon Kugelhopfs, Walnut Kugelhopfs, and the Exquisite "Kugelhopf Surprise"

In recent times several other types of kugelhopfs have been developed. These include bacon kugelhopfs, walnut kugelhopfs, and the "kugelhopf surprise," all of which are especially tempting gastronomic creations.

■

The regional brioche recipes might seem somewhat obscure at first glance and with the exception of the *kugelhopf*, are seldom encountered. It is also certain that each region remains loyal to its traditional brioche to the exclusion of those of neighboring provinces. However, this chapter remains a source for scholars, fertile ground for curious bakers, and is uniquely useful for anyone interested in regional baking.

Figure 14–1 Kugelhopf.

Exhibit 14–1 The Traditional Kugelhopf by Sponge and Dough Method

Formula

	Weight		Yeast Sponge	Final Dough
Strong flour	2 kg (70.55 oz)	100%	480 g (16.93 oz)	1.52 kg (53.62 oz)
Salt	40 g (1.41 oz)	2%		40 g (1.41 oz)
Yeast	80 g (2.82 oz)	4%	8 g (0.28 oz)	72 g (2.54 oz)
Sugar	440 g (15.52 oz)	22%		440 g (15.52 oz)
Butter*	700 g (24.69 oz)	35%		700 g (24.69 oz)
Eggs	640 g (22.57 oz)	32%		640 g (22.57 oz)
Milk	640 g (22.57 oz)	32%	290 g (10.23 oz)	340 g (11.99 oz)
Raisins	400 g (14.11 oz)	20%		400 g (14.11 oz)
Schnapps	60 g (2.12 oz)	3%		60 g (2.12 oz)
Orange flower water	80 g (2.82 oz)	4%		80 g (2.82 oz)
Ascorbic acid	(20 mg)	(10 ppm)		(50 mg)
YEAST SPONGE			778 g (27.44 oz)	778 g (27.44 oz)
TOTAL DOUGH WEIGHT				5.08 kg (179.12 oz)
Almonds (to be placed in bottom of molds–not in dough)	120 g (4.23 oz)	6%		120 g (4.23 oz)

Procedures

			Yeast Sponge	Final Dough
Mixing: Low speed			6 min	4 min
Beater: Second speed				10 min
Dough temperature from mixer			(25° C / 77° F)	(26° C / 78.8° F)
Rest period for yeast sponge			(3 h @ 25° C / 77° F *or* 16 h @ +4° C / 39.2° F)	
First fermentation				160 min
Degas (punch down) after				(100 min)
Division, rounding & rest period				40 min
Dough forming				
Second fermentation				100 min
Baking (@ 210° C / 410° F)				35 min
PRODUCTION TIME				6 h

*If salted butter must be used for the recipes in this section, reduce the quantity of salt accordingly (see Chapter 13).

Exhibit 14–2 Formula and Procedure for Straight Dough Kugelhopf Hearth Cake

Formula

Strong flour	2 kg (70.54 oz)	100%
Salt	40 g (1.41 oz)	2%
Yeast	90 g (3.17 oz)	4.5%
Sugar	440 g (15.52 oz)	22%
Butter	640 g (22.57 oz)	32%
Eggs	640 g (22.57 oz)	32%
Milk	700 g (24.69 oz)	35%
Raisins	400 g (14.11 oz)	20%
Schnapps	70 g (2.47 oz)	3.5%
Orange flower water	80 g (2.82 oz)	4%
Ascorbic acid	(30 mg)	(15 ppm)
TOTAL DOUGH WEIGHT	5.1 kg (179.89 oz)	
Almond halves (Place in bottom of mold— do not add to dough)	120 g (4.23 oz)	6%

Procedures

Mixing: 1st speed	4 min
Beater: 2nd speed	10 min
Dough temperature	(25° C / 77° F)
1st fermentation	200 min
Degassing (after)	(90 min)
Divide, round, and rest	30 min
Dough forming	
2nd fermentation	100 min
Baking (210° C / 410° F)	35 min
TOTAL TIME	6 h 20 min

It is advisable to halve the normal sugar content in making the bacon kugelhopf. The raisins are replaced by small cubes of "streaky" bacon browned in a frying pan, then drained and cooled before being incorporated into the dough.[3] In the case of the walnut kugelhopf, the raisins are replaced by an equal or slightly greater quantity of coarsely chopped walnuts, and the butter content of the formula is reduced somewhat.

The "kugelhopf surprise" is made according to the traditional formula, with a slight reduction in sugar. After a 24-hour storage period, the kugelhopf is sliced both vertically and horizontally in order to obtain small slices about 6 mm thick. These are buttered and made into small sandwiches, using as fillings foie gras, duck, ham, or cheeses such as Comté from the Jura region or light goat milk cheese. When these preparations have been completed, the kugelhopf is reassembled. As such, it is certainly a wonderful culinary delicacy.

It is interesting to note that Alsatian bakers and pastry chefs took the important step in 1976 of creating a quality label under the designation "Kugelhopf Alsace" in order to combat the degradation of the quality and originality of the authentic Alsatian kugelhopf. Both the formula and methods of production of this fine product are very close to the traditional kugelhopf formulas given in this book.

Two general observations should be made regarding the occasional use of dough improvers and the proofing of formed hearth cakes (brioches):

a) It may be advisable, for all of the following regional brioche formulas, to improve the physical properties of the doughs by adding ascorbic acid. This will have the effect of improving their handling qualities and their shape and appearance, as well as providing better volume after baking. Use the following guidelines to avoid excessive use of this dough conditioner:

- *5 to 10 ppm for formulas using natural sponge leavening or natural dough leavening (levain de pâte);*
- *10 to 15 ppm for formulas using the sponge and dough system, or 10 to 15 mg per kilogram of flour;*
- *15 to 20 ppm for straight dough formulas.*

b) Placing the formed hearth cakes into a proof cabinet maintained at 28 to 30°C allows the baker to considerably reduce the second fermentation period. However, care must be taken to maintain the relative humidity of the proof chamber at around 80%.

VENDÉE-STYLE BRIOCHE

The Vendée-style brioche of the past was of good quality, and it could continue to be even today. Just as all regional brioches, it is higher in sugar content and contains less butter than the traditional French brioche, typified by the Paris type. In the old days, it was always leavened with a natural sponge. It has commonly been made that way since, and in a sense often still is, since sponge and dough production is common. However, over the past several decades its production has changed a great deal, and today it is often made by the straight dough method.

In addition, the dough is usually made with the addition of emulsifiers—generally esterified mono- and diglycerides of fatty acids. These additives, in combination with ascorbic acid and the practice of high-speed mixing, serve to increase the volume and lightness of the brioche, but the resulting crumb is excessively homogeneous and has an unpleasant mouthfeel, much like cotton wool.

This type of brioche is generally machine-shaped in the form of a braid, or twisted and baked in a baking mold (Figure 14–2). The mechanical forming of the braid only serves to accentuate the cotton-wool texture and unpleasant mouthfeel of the crumb.

This is the typical Vendée-style brioche, as it is most often produced on a large industrial scale. This situation is not universal, but regrettably, it is all too widespread. However, good quality Vendée brioches are still to be found. The type of formula in Exhibit 14-3 respects the traditional practice of sponge and dough production and allows the baker to make a brioche with excellent keeping qualities, fine flavor, and good softness.

It should be noted that the formula percentage of sugar, butter, and eggs may vary, and increases in the level of these ingredients—especially the first—make it necessary to also increase the percentage of yeast. In addition, the baker should know that using milk

Figure 14–2 Gâche Vendéeme and Brioche Vendéeme (Braided Loaf).

as a flour wetting agent produces a slower rise than when the flour is hydrated with water.

Gâche Hearth Cake

Another type of brioche, which is only very slightly different from the preceding, is also encountered in the Vendée. It is commonly known as the "baker's peel" hearth cake or *gâche*. Its chief distinction is that it is made only in the shape of a shuttlecock (weaver's shuttle), which is given a single lengthwise razor slash before being placed in the oven. The formula is a little higher in sugar and a little lower in butter and eggs. It is also most commonly made by the straight dough method, even though the sponge and dough system would produce better results. For that reason the formula in Exhibit 14-4 includes a prefermented culture.

Exhibit 14–3 A Quality Vendée-Style Brioche

Formula

	Weight		Yeast Sponge	Final Dough
Flour	2 kg (70.55 oz)		600 g (21.16 oz)	1400 g (49.38 oz)
Salt	40 g (1.41 oz)	2%	6 g (0.21 oz)	34 g (1.2 oz)
Yeast	70 g (2.47 oz)	3.5%	8 g (0.28 oz)	62 g (2.19 oz)
Milk	400 g (14.11 oz)	20%	400 g (14.11 oz)	
Eggs	710 g (25.04 oz)	35.5%		710 g (25.04 oz)
Sugar	500 g (17.64 oz)	25%		500 g (17.64 oz)
Butter	520 g (18.34 oz)	26%		520 g (18.34 oz)
Orange flower water	50 g (1.76 oz)	2.5%		50 g (1.76 oz)
Ascorbic acid	(100 mg)	(10 ppm)		(100 mg)
YEAST SPONGE			1014 g (35.77 oz)	1014 g (35.77 oz)
TOTAL DOUGH WEIGHT				4.29 kg (151.32 oz)

Procedures

Mixing: Low speed			4 min	4 min
Beater: Second speed			4 min	12 min
Dough temperature from mixer			(25° C / 77° F)	(25° C / 77° F)
Rest period for yeast sponge			5 h @ 24° / 77° F C	
First fermentation				180 min
Punch down (degassing) after				(110) min
Dough division & rest				25 min
Dough makeup				
Second fermentation				160 min
Baking @ 210° C				35–40 min
PRODUCTION TIME				7 h

Exhibit 14–4 The Gâche (Vendée-Style Hearth Cake)

Formula

	Weight		Yeast Sponge with Milk	Final Dough
Strong flour	2 kg (70.55 oz)		500 g (17.64 oz) (25%)	1.5 kg (52.9 oz)
Salt	36 g (1.27 oz)	1.8%	8 g (0.28 oz)	28 g (0.99 oz)
Yeast	80 g (2.82 oz)	4%	5 g (0.18 oz)	75 g (2.65 oz)
Milk	330 g (11.64 oz)	16.5%	330 g (11.64 oz)	
Eggs	640 g (22.57 oz)	32%		640 g (22.57 oz)
Sugar	560 g (19.75 oz)	28%		560 g (19.75 oz)
Butter	300 g (10.58 oz)	15%		300 g (10.58 oz)
Dairy cream	160 g (5.64 oz)	8%		160 g (5.64 oz)
Orange flower water	60 g (2.12 oz)	3%		60 g (2.12 oz)
Ascorbic acid	(20 mg)	(10 ppm)		(20 mg)
YEAST SPONGE			843 g (29.74 oz)	843 g (29.74 oz)
TOTAL DOUGH WEIGHT				4166 g or 4.166 kg (146.95 oz)

Procedures

Mixing: Low speed			4 min	4 min
Beater: Second speed			4 min	12 min
Dough temperature from mixer			(25° C / 77° F)	(25° C / 77° F)
Rest period for yeast sponge			5 h @ 25° C / 77° F	
First fermentation				210 min
Punch down (degassing) after				(150 min)
Dough division & rest				25 min
Dough makeup				
Second fermentation				200 min
Baking @ 200° C / 392° F				30–40 min
PRODUCTION TIME				8 h 15 min

SPECIALTY BRIOCHES

Gâteau des Rois

This holiday product, also known as 12th night cake or Epiphany cake, is generally made in the Bordeaux and Lyon regions, and southern France in general, around January 6, which is the traditional holy day of the Epiphany or Feast of the Three Kings. The dough used in this version of the Kings' cake is closely related to brioche dough. It has a high sugar content and may have a slightly lower butter or fat content around Bordeaux and in southwest France.

Kings' cake is made in a ring or crown shape and is brushed with a whole egg wash before baking (Figure 14-3). In Bordeaux it is sprinkled with white pearl sugar, while in Lyon it is snipped with a scissors to form a sawtooth pattern.

The tradition of a Kings' cake made from rich brioche dough is also observed in the Romanche area of Switzerland, but in that region it is normally baked in a round layer cake pan. The dough is divided into nine portions, one of which is slightly larger than the others. The individual dough pieces are rounded, and the largest dough ball is placed in the center of the baking tin, surrounded by the eight others. The effect created is that of a large flower head surrounded by eight petals. The proofed cake is brushed with egg wash and then covered with blanched almonds before baking.

Two formulas for the production of these cakes are given, one of which uses the straight dough method (Exhibit 14-5), the other the sponge and dough technique (Exhibit 14-6). In both these cases, the baker may vary the egg, butter, and sugar levels to change the richness and quality of the final product.

Fouace des Rameaux

Palm Sunday precedes Easter, which may occur from the end of March to mid-April. Thus, Palm Sunday may arrive as early as the beginning of spring, or as late as a fortnight or even a month later. In either case, it follows the long winter and coincides with the reawakening of nature. The days have lengthened, the sun has warmed the earth, and vegetation bursts forth. It is also the time when hens begin to lay, and eggs once again become abundant.

In the wine-producing regions even today, it is the time when wine is drawn for bottling. White wine is drawn off and filtered, and the clarified wine is bottled by the winemaker. In the case of those wines that contain a little sugar and some active yeasts, spring is also the time that they reawaken and become lively, foamy, and seductively enticing when uncorked.

Figure 14-3 (A) Gâteau des Rois de Lyon, (B) Fouace des Rameaux, (C) Pogne de Romans.

Exhibit 14–5 Formula and Procedure for Straight Dough Kings' Cake

Formula

Strong flour	2 kg (70.55 oz)	100%
Salt	40 g (1.41 oz)	2%
Yeast	70 g (2.47 oz)	3.5%
Sugar	440 g (15.52 oz)	22%
Butter	400 g (14.11 oz)	20%
Eggs	640 g (22.57 oz)	32%
Milk	440 g (15.52 oz)	22%
Orange flower water	50 g (1.76 oz)	2.5%
Ascorbic acid	(50 mg)	(10 ppm)
TOTAL DOUGH WEIGHT	4080 g or 4.08 kg (143.92 oz)	

Procedures

Mixing: 1st speed	4 min
Beater: 2nd speed	12 min
Dough temperature	(25° C / 77° F)
1st fermentation	200 min
Degassing (after)	(120 min)
Divide form	25 min
2nd fermentation	150 min
Baking (210° C / 410° F)	35 min
TOTAL TIME	7 h

These two happy coincidences have given rise to a tradition of baking a special Palm Sunday hearth cake, which is an excellent companion to the white wine of the previous autumn. This baking was commonly done in rural households and even in the cities of lower Languedoc, in the southwest of France, in other regions, and even in some foreign countries. Thus, these crowning jewels of the process of alcoholic fermentation were united in a delicate gastronomic harmony.

Professional bakers made these hearth cakes for households that did not produce them, but by tradition most housewives made them. The housewife obtained a bit of "mother dough" from the baker, made up the mother sponge, oversaw the proper mixing of the dough, carefully observed the first fermentation, and laboriously made up the hearth cakes. She then carried them to the bakery. The baker oversaw the second fermentation, and at the proper time baked and finished the product.

In hamlets where one of the homes had a bread oven, this tradition was the impetus for putting it into order, firing it up, and baking several oven loads

---

Exhibit 14–6 Formula and Procedure for Kings' Cake by Sponge and Dough Method

Formula

	Weight		Yeast Sponge	Final Dough
Strong flour	2 kg (70.55 oz)		480 g (16.93 oz) (24%)	1.52 kg (53.62 oz)
Salt	40 g (1.41 oz)	2%		40 g (1.41 oz)
Yeast	70 g (2.47 oz)	3.5%	12 g (0.42 oz)	58 g (2.05 oz)
Sugar	480 g (16.93 oz)	24%		480 g (16.93 oz)
Butter	400 g (14.11 oz)	20%		400 g (14.11 oz)
Eggs	700 g (24.69 oz)	35%		700 g (24.69 oz)
Milk	360 g (12.7 oz)	18%	288 g (10.16 oz)	72 g (2.54 oz)
Orange flower water	50 g (1.76 oz)	2.5%		50 g (1.76 oz)
Ascorbic acid	(50 mg)	(10 ppm)		(50 mg)
YEAST SPONGE			780 g (27.51 oz)	780 g (27.51 oz)
TOTAL DOUGH WEIGHT				4100 g or 4.1 kg (144.62 oz)

Procedures

Mixing: Low speed		7 min	4 min
Beater: Second speed			12 min
Dough temperature from mixer		(24° C / 75° F)	(25° C / 77° F)
Rest period for yeast sponge		3 h @ 25° C / 77° F *or* 15 h @ +4° C / 39° F	
First fermentation			160 min
Dough division & forming			35 min
Second fermentation			140 min
Baking @ 200° C / 392° F			35 min
PRODUCTION TIME			6 h 26 min

for the entire village. Whether the hearth cakes were baked in the baker's oven or on a domestic hearth, this was always the time for comparison and comments on the respective merits of the hearth cakes by the women involved, and also the occasion of much good-natured gossip.

This production was generally sufficient to supply several hearth cakes to each household. In this wine-producing region, the consumption of the hearth cakes was always accompanied by the drinking of white wine from the previous season. This generally went on for around 10 days without the hearth cakes losing their flavor and their other attractive qualities, because of the practice of leavening them with a natural yeast sponge.

Two formulas are given below for the Palm Sunday cake, one made with a naturally fermented sponge (Exhibit 14–7), the other with the sponge and dough method (Exhibit 14–8). As I noted previously, the preparation and the cultivation of the naturally fermented

Exhibit 14–7 Naturally Fermented Sponge for Fouace des Rameaux

Rafraîchi (Rebuilding Culture)		*Naturally Fermented Sponge*	
Mother dough (chef)*	80 g (2.82 oz)	Rebuilt culture	266 g (9.38 oz)
Flour	120 g (4.23 oz)	Flour	300 g (10.58 oz)
Water	66 g (2.32 oz)	Milk	174 g (6.14 oz)
Dough temperature from mixer	27° C / 80.6° F	Dough temperature from mixer	27° C / 80.6° F
TOTAL	266 g (9.38 oz)	TOTAL	740 g (26.10 oz)
FERMENTATION TIME	4 h approx.	FERMENTATION TIME (minimum)	5 h

Fouace des Rameaux from Naturally Fermented Sponge

Formula

	Weight		*Natural Sponge*	*Final Dough*
Strong flour	2 kg (70.55 oz)	100%	420 g (14.81 oz) (23.5%)	1580 g (55.73 oz)
Salt	40 g (1.41 oz)	2%		40 g (1.41 oz)
Yeast	4 g (0.14 oz)	0.2%		4 g (0.14 oz)
Sugar	500 g (17.64 oz)	25%		500 g (17.64 oz)
Butter	300 g (10.58 oz)	15%		300 g (10.58 oz)
Dairy cream	200 g (7.05 oz)	10%		200 g (7.05 oz)
Eggs	600 g (21.16 oz)	30%		600 g (21.16 oz)
Milk	400 g (14.11 oz)	20%	174 g (6.14 oz)	226 g (7.97 oz)
Water	66 g (2.33 oz)	3.3%	66 g (2.33 oz)	
Orange flower water	60 g (2.12 oz)	2.5%		60 g (2.12 oz)
Candied peel	300 g (10.58 oz)	15%		300 g (10.58 oz)
Chef			80 g (2.82 oz), (50 flour, 30 water)	80 g (2.82 oz)
NATURAL SPONGE			740 g (26.10 oz)	740 g (26.10 oz)†
TOTAL DOUGH WEIGHT				4.55 kg or 4550 g (160.47 oz)

Procedures

Mixing: Low speed		6 min	4 min
Beater: Second speed			10 min
Dough temperature from mixer		(27° C / 80.6° F)	(26° C / 78.8° F)
Fermentation period for natural sponge		5 h @ 24° C / 75.2° F	
First fermentation			5 to 7 h
Dough division & rest period			40 min
Dough forming			
Second fermentation			3 h 30 min to 4 h
Baking @ 210° C / 410° F			40 min
PRODUCTION TIME			

*The "chef" or mother dough is made up of around 50 g of flour and 30 g of water.
†Weight of total natural sponge is 47% of the weight of flour used in the final dough, or 1580 x .47 = 742.6 g).

Exhibit 14–8 Fouace des Rameaux by Sponge and Dough Method

Formula

	Weight		Yeast Sponge	Final Dough
Strong flour	2 kg (70.55 oz)	100%	400 g (14.11 oz) (20%)	1.6 kg (56.44 oz)
Salt	40 g (1.41 oz)	2%		40 g (1.41 oz)
Yeast	70 g (2.47 oz)	3.5%	6 g (0.21 oz)	64 g (2.26 oz)
Sugar	480 g (16.93 oz)	24%		480 g (16.93 oz)
Butter	400 g (14.11 oz)	20%		400 g (14.11 oz)
Dairy cream	100 g* (3.53 oz)	5%		100 g (3.53 oz)
Eggs	700 g (24.69 oz)	35%		700 g (24.69 oz)
Milk	340 g (11.99 oz)	17%	260 g (9.17 oz)	80 g (2.82 oz)
Candied peel	340 g (11.99 oz)	15%		340 g (11.99 oz)
Almond halves	120 g (4.23 oz)	6%		120 g (4.23 oz)
Orange flower water	50 g (1.76 oz)	2.5%		50 g (1.76 oz)
Ascorbic acid	(20 mg)	(10 ppm)		(20 mg)
YEAST SPONGE			666 g (23.49 oz)	666 g (23.49 oz)
TOTAL DOUGH WEIGHT				4.64 kg (163.67 oz)

Procedures

		Yeast Sponge	Final Dough
Mixing: Low speed		6 min	4 min
Beater: Second speed			12 min
Dough temperature from mixer		(25° C / 77° F)	(25° C / 77° F)
Sponge rest period		4 h @ 24° C / 75.2° F or 16 h @ +5 ° C / 41° F	
First fermentation			180 min
Punch down (degas) after			(100 min)
Dough division & rest period			40 min
Dough forming			
Second fermentation			120 min
Baking @ 200° C / 392° F			35 min
PRODUCTION TIME			6 h 30 min

*The original formula gave 850 g in error for this ingredient.

---

sponge produce a culture which may be adapted for use in the production of other bakery goods of this type.

The dough used for the Palm Sunday hearth cake has a relatively high sugar content and a moderately high butter content, with the addition of fermented dairy cream. This cream should be more mature (aged to a greater degree) than that which is used for whipped cream.[4]

In the past, Palm Sunday hearth cake often contained preserved fruit (such as candied lemon, orange, or citron peel or melon rind), and the crown was decorated with slices of candied melon. Alternatively, it was dusted with pearl sugar and decorated with almond halves before baking. These practices are not so common today.

Fouace de Rodez

The hearth cake made in Rodez is similar in many ways to the Palm Sunday cake. It contains only very slightly differing amounts of eggs, sugar, and butter. As in the Palm Sunday cake, part of the butter can be replaced by well-fermented dairy cream, which is called "fresh cream" in France.[5] The greater acidity contributed by the cream improves both the taste and keeping qualities of the product.

The Rodez hearth cake is only very rarely made with candied fruit. As in the case of the Palm Sunday cake, it was formerly made from a naturally leavened sponge. Today, however, it is most often made by the straight dough method (Exhibit 14–9), and natural sponge leavening (Exhibit 14–10) or *levain de pâte* (a

Exhibit 14–9 Formula and Procedure for Straight Dough Fouace de Rodez (Hearth Cake)

Formula

Strong flour	2 kg (70.55 oz)	100%
Salt	40 g (1.41 oz)	2%
Yeast	80 g (2.82 oz)	4%
Sugar	520 g (18.34 oz)	26%
Butter	460 g (16.23 oz)	23%
Dairy cream	140 g (4.94 oz)	7%
Eggs	680 g (23.99 oz)	34%
Milk	360 g (12.7 oz)	18%
Wine vinegar*	24 g (0.84 oz)	1.2%
Orange flower water	70 g (2.47 oz)	3.5%
Ascorbic acid	(20 mg)	(20 ppm)
TOTAL DOUGH WEIGHT	4374 g (154.29 oz)	

Procedures

Mixing: 1st speed	4 min
Beater: 2nd speed	10 min
Dough temperature	(25° C / 77° F)
1st fermentation	5 h
Degassing (after)	(3 h) & (4 h)
Divide, round and rest	40 min
Dough forming	
2nd fermentation	3 h
Baking (210° C / 410° F)	30–40 min
TOTAL TIME	9 h 35 min

*The addition of a small amount of wine vinegar raises the acidity of the dough slightly and improves the keeping qualities of the hearth cake.

---

stiff natural sponge with added baker's yeast, Exhibit 14–11) are used more and more rarely. This is unfortunate, considering the quality and distinctive product characteristics that result from the use of both these traditional leavening methods.

Many professional bakers assume that the straight dough method allows them to save time in the production of these and other products, whereas a properly scheduled natural sponge with added baker's yeast only requires a slight increase in the production work required. This increase is unimportant in comparison to the resulting improvements in overall product quality.

La Pogne de Romans

The "fist" of Romans (*la pogne de Romans*) is a brioche or hearth cake that may be encountered from the south of Lyon to the Mediterranean coast. This product has an especially interesting name, which originated during the Middle Ages. During the 14th and 15th centuries, wheat was grown only very rarely in this southeastern part of France, and bread was made exclusively from rye, the only cereal grain that was commonly raised in the region at that time.

It was the custom in the area around Romans-sur-Isère to make a yeast-leavened holiday cake containing eggs and rye flour. However, to improve both the taste and the appearance of the cake, the baker added several "fistfuls" of wheat flour. This is how the "fist" name came about, and how it took its place in the culinary vocabulary of the Romans area.

This product is made in the form of a crown or ring (See Figure 14–3) and is sometimes quite large. It is often made from a naturally leavened sponge even today, or even more often from a naturally leavened dough to which has been added a small amount of baker's yeast in order to produce a lighter cake (Exhibits 14–12 and 14–13). Unfortunately, the dough is often overmixed to obtain a larger volume, which has the effect of reducing the outstanding flavor, taste, and eating qualities of a hearth cake that would otherwise be excellent.

This is one of the most flavorful and highly scented of all the regional hearth cakes, doubtless because of the influence of the cooks and bakers of Provence, which is nearby. When the "pogne" is made by housewives, it is customary to add a small amount of an aperitif made from sweet Muscat raisins. The crown may also be decorated by placing several red almond pralines on the upper surface during the forming stage. Crowns that are brushed with whole egg wash may also be sprinkled with a little pearl sugar before baking.

Mouna

The *mouna* is a brioche or hearth cake with a high sugar content. It is Spanish in origin, but is commonly made in Morocco and Algeria as well. It is sometimes also found in France, where it really deserves to be

Exhibit 14–10 Naturally Fermented Sponge for Fouace de Rodez

Rafraîchi (Rebuilding Culture)		Naturally Fermented Sponge (Levain)	
Mother dough (chef)*	80 g (2.82 oz)	Rebuilt culture	266 g (9.38 oz)
Flour	120 g (4.23 oz)	Flour	300 g (10.58 oz)
Water	66 g (2.33 oz)	Milk	174 g (6.14 oz)
Dough temperature from mixer	27° C / 80.6° F	Dough temperature from mixer	27° C / 80.6° F
TOTAL	266 g (9.38 oz)	TOTAL	740 g (26.1 oz)
FERMENTATION TIME	4 h approx.	FERMENTATION TIME (minimum)	5 h

Fouace de Rodez from Naturally-Fermented Sponge

Formula

(Note: Fermentation levels noted at the end of 1st and 2nd fermentation periods are given in relation to starting dough volume.)

	Weight		Natural Sponge	Final Dough
Strong flour	2 kg (70.55 oz)	100%	420 g (14.81 oz) (29.5%)	1.58 kg (55.73 oz)
Salt	30 g (1.06 oz)	1.5%		30 g (1.06 oz)
Sugar	500 g (17.64 oz)	25%		500 g (17.64 oz)
Butter	500 g (17.64 oz)	15%		500 g (17.64 oz)
Eggs	600 g (21.16 oz)	30%		600 g (21.16 oz)
Milk	360 g (12.7 oz)	15%	174 g (6.14 oz)	186 g (6.56 oz)
Orange flower water	50 g (1.76 oz)	2.5%		50 g (1.76 oz)
Water	66 g (2.33 oz)	4.5%	66 g (2.33 oz)	
Orange essence	(4 drops)			(4 drops)
Chef			80 g (2.82 oz) (50 flour + 30 water)	
NATURAL SPONGE			740 g (26.10 oz)	740 g (26.1 oz)†
TOTAL DOUGH WEIGHT				4186 g (147.65 oz)

Procedures

Mixing: Low speed		6 min	4 min
Beater: Second speed			10 min
Dough temperature from mixer		(26° C / 78.8° F)	(26° C / 78.8° F)
Fermentation period for natural sponge		5 h @ 24° C / 75.2° F	
First fermentation (2.5 x original vol.)			10 to 12 h
Division, rounding & rest period			40 min
Dough forming			
Second fermentation (3.5 x original vol.)			4 to 5 h
Baking @ 210° C / 410° F			35 min
PRODUCTION TIME			18 h 30 min

*The "chef" or mother dough is made up of around 50 g of flour and 30 g of water.
†Weight of total natural sponge is 47% of the weight of flour used in the final dough, or 1580 x .47 = 742.6 g.

better known. It is sometimes flavored with anise, but since it is nearly always flavored with orange flower water, essence of orange, and perhaps even candied orange peel, its most notable aroma combines all the rich perfumes of the orange tree, which is above all others the fruit tree of these Mediterranean lands.

In Spain the *mouna* was the feast day cake made during the Easter season, just as the Palm Sunday

Exhibit 14–11 Levain de Pâte (Mixed Method: Naturally Fermented Sponge with Baker's Yeast) for Rodez Hearth Cake*

Rebuilding Culture (Rafraîchi)		*Naturally Fermented Dough*	
Mother dough (chef)†	60 g (2.12 oz)	Rebuilt culture	207 g (7.30 oz)
Flour	92 g (3.25 oz)	Flour	220 g (7.76 oz)
Water	55 g (1.94 oz)	Milk	130 g (4.59 oz)
Dough temperature from mixer	26° C / 78.8° F	Dough temperature from mixer	26° C / 78.8° F
TOTAL	207 g (7.30 oz)	TOTAL	557 g (19.65 oz)
FERMENTATION TIME	4 h minimum	FERMENTATION TIME (minimum)	4 h

Rodez Hearth Cake from Naturally-Prefermented Dough

Formula

	Weight		*Natural Dough*	*Final Dough*
Strong flour	2 kg (70.55 oz)	100%	350 g (12.35 oz) (17.5%)	1650 g (58.20 oz)
Salt	30 g (1.06 oz)	1.5%		30 g (1.06 oz)
Yeast	20 g (0.71 oz)	1%		20 g (0.71 oz)
Sugar	500 g (17.64 oz)	25%		500 g (17.64 oz)
Butter	440 g (15.52 oz)	22%		440 g (15.52 oz)
Dairy cream	160 g (5.64 oz)	8%		160 g (5.64 oz)
Eggs	640 g (22.57 oz)	32%		640 g (22.57 oz)
Milk	300 g (10.58 oz)	12%	280 g (9.88 oz)	170 g (5.99 oz)
Water	55 g (1.94 oz)	3.5%	175 g (6.17 oz)	
Orange flower water	60 g (2.12 oz)	3%		60 g (2.12 oz)
Lemon essence	(10 drops)		—	(10 drops)
Chef			60 g (2.12 oz) (38 flour + 22 water)	60 g (2.12 oz)
NATURAL PREFERMENT				595 g (20.99 oz)
TOTAL DOUGH WEIGHT			595 g (20.99 oz)	4265 g (150.44 oz)

Procedures

Mixing: Low speed		6 min	4 min
Beater: Second speed			10 min
Dough temperature from mixer		(26° C / 78.8° F)	(26° C / 78.8° F)
Fermentation period for natural sponge		5 h @ 24° C / 75.2° F	
First fermentation (3 x original vol.)			6 to 8 h
Dough division & rest period			40 min
Dough forming			
Second fermentation (3.5 x original vol.)			4 h
Baking @ 210° C / 410° F			35 min
PRODUCTION TIME			13 h 30 min

*The term *levain de pâte* is used for this method, even though a small amount of baker's yeast is used in the final dough.
†The "chef" or mother dough is made up of around 38 g (1.34 oz) of flour and 22 g (0.77 oz) of water.
Flour used in natural dough is 32.25 parts of the 100 parts of flour used in the total dough.

hearth cake has always been in southeastern France when eggs were readily available. The *mouna* was generally made in the form of a crown or ring, although it was also produced as a ball-shaped loaf. By tradition, an unbroken egg—shell and all—(the Easter egg) was pressed into the dough and baked with the hearth cake. This added nothing to the flavor and is no longer practiced either in Spain or in North Africa.[6]

Exhibit 14–12 Naturally Prefermented Dough for "Pogne" de Romans-sur-Isère*

Rebuilding Culture		Naturally Prefermented Dough	
Mother dough (chef)†	60 g (2.12 oz)	Rebuilt culture	207 g (7.30 oz)
Flour	92 g (3.25 oz)	Flour	220 g (7.05 oz)
Water	55 g (1.94 oz)	Milk	130 g (4.59 oz)
Dough temperature from mixer	26° C / 78.8° F	Dough temperature from mixer	26° C / 78.8° F
TOTAL	207 g (7.30 oz)	TOTAL	557 g (19.65 oz)
FERMENTATION TIME	4 h minimum	FERMENTATION TIME (minimum)	4 h

"Pogne" de Romans' from Naturally-Prefermented Dough

Formula

	Weight		Natural Dough	Final Dough
Strong flour	2 kg (70.55 oz)		350 g (12.35 oz) (17.5%)	1.65 kg (58.20 oz)
Salt	36 g (1.27 oz)	1.8%		36 g (1.27 oz)
Yeast	20 g (0.71 oz)	1%		20 g (0.71 oz)
Sugar	540 g (19.05 oz)	27%		540 g (19.05 oz)
Butter	500 g (17.63 oz)	25%		500 g (17.64 oz)
Eggs	720 g (25.40 oz)	36%		720 g (25.4 oz)
Milk	300 g (10.58 oz)	15%	130 g (4.59 oz)	170 g (6.0 oz)
Water	55 g (1.94 oz)	2.8%	55 g (1.94 oz)	
Orange flower water	100 g (3.53 oz)	5%		100 g (3.53 oz)
Lemon essence	(4 drops)			(4 drops)
Chef			60 g (2.12 oz) (38 flour + 22 water)	60 g (2.12 oz)
NATURAL PREFERMENT			595 g (20.99 oz)	595 g (20.99 oz)
TOTAL DOUGH WEIGHT				4331 g (152.77 oz)

Procedures

Mixing: Low speed		6 min	4 min
Beater: Second speed			10 min
Dough temperature from mixer		(26° C / 78.8° F)	(25° C / 78.8° F)
Fermentation period for natural sponge		5 h @ 25° C / 77° F	
First fermentation			6 to 7 h
Dough division & rest period			40 min
Dough forming			
Second fermentation			4 h
Baking @ 200° C / 392° F			35 min
PRODUCTION TIME			12 h 30 min

*This term is used for this method, even though a small amount of baker's yeast is used in the final dough.
†The "chef" or mother dough is made up of around 38 g (1.34 oz) of flour and 22 g (0.78 oz) of water.
Flour used in natural dough is 32.25 parts of the 100 parts of flour used in the total dough.

The *mouna* may be made from a straight dough, but as with the preceding hearth cakes, the use of the sponge and dough method improves the quality to a noticeable degree—which is why the formula given in Exhibit 14–14 uses the sponge and dough technique.

Pompe des Rois (a 12th Night Variant)

This is a product of the south-central parts of France, the Cévennes region. It is a brioche or hearth cake that is made not only for the Feast of the Three

Exhibit 14–13 "Pogne de Romans" by Sponge and Dough Method

Formula

	Weight		Yeast Sponge	Final Dough
Strong flour	2 kg (70.55 oz)		500 g (17.64 oz) (25%)	1.5 kg (52.91 oz)
Salt	36 g (1.27 oz)	1.8%		36 g (1.27 oz)
Yeast	70 g (2.47 oz)	3.5%	6 g (0.21 oz)	64 g (2.26 oz)
Sugar	500 g (17.64 oz)	25%		500 g (17.64 oz)
Butter	600 g (21.16 oz)	30%		600 g (21.16 oz)
Eggs	700 g (24.70 oz)	35%		700 g (24.70 oz)
Milk	300 g (10.58 oz)	15%	300 g (10.58 oz)	
Orange flower water	100 g (3.53 oz)	5%		100 g (3.53 oz)
Wine vinegar	24 g (0.84 oz)	1.2%		24 g (0.84 oz)
Lemon essence	(4 drops)			(4 drops)
Ascorbic acid	(20 mg)	(10 ppm)		(20 mg)
YEAST SPONGE			806 g (28.43 oz)	806 g (28.43 oz)
TOTAL DOUGH WEIGHT				4.33 kg (152.73 oz)

Procedures

		Yeast Sponge	Final Dough
Mixing: Low speed		6 min	4 min
Beater: Second speed			10 min
Dough temperature from mixer		(25° C / 77° F)	(25° C / 77° F)
Rest period for yeast sponge		5 h @ 24° C / 75.2° F or 15 h @ + 4° C / 39.2° F	
First fermentation			4 h
Degas (punch down) after			(2 h 30 min)
Division, rounding & rest period			40 min
Dough forming			
Second fermentation			2 h 30 min
Baking @ 210° C / 410° F			35 min
PRODUCTION TIME			8 h

Kings, but also during the months which precede or follow it. It is rich, with a pleasant aroma and excellent keeping qualities, so here again the use of the sponge and dough method is most advisable (Exhibit 14–15).

The mixing technique is identical to that used for all brioche (hearth cake) doughs: the raisins and candied fruit are incorporated at the end of the mixing cycle, using only first (low) speed. The dough is formed into "crowns" or rings first by being formed into a ball. The dough ball is then stretched and flattened slightly and pierced in the center with a hole, which is enlarged as desired.

Before baking, the crowns are brushed with whole egg wash and sprinkled with the pearl sugar noted in the formula. Strips of candied fruits—melon, angelique, or orange—may also be used to decorate the crowns before baking.

Gibassier

The formula for the gibassier brioche, which resembles a sand dollar, is typical of Provence, with the native aromas of the region: olive oil, anise, and orange flowers (Exhibit 14–16). The usual leavening

Exhibit 14–14 "Mouna" Spanish Hearth Cake by Sponge and Dough Method

Formula

	Weight		Yeast Sponge	Final Dough
Strong flour	2 kg (70.55 oz)	100%	600 g (21.16 oz) (30%)	1.4 kg (49.38 oz)
Salt	36 g (1.27 oz)	1.8%		36 g (1.27 oz)
Yeast	90 g (3.17 oz)	4.5%	12 g (0.42 oz)	78 g (2.75 oz)
Sugar	560 g (19.75 oz)	28%		560 g (19.75 oz)
Butter	640 g (22.57 oz)	32%		640 g (22.57 oz)
Eggs*	740 g (26.10 oz)	36%		740 g (26.10 oz)
Milk	340 g (11.99 oz)	17%	340 g (11.99 oz)	
Orange flower water	120 g (4.23 oz)	6%		120 g (4.23 oz)
Orange peel	320 g (11.29 oz)	16%		320 g (11.29 oz)
Orange essence	(4 drops)			(4 drops)
Ascorbic acid	(20 mg)	(20 ppm)		(20 mg)
YEAST SPONGE			952 g (33.58 oz)	952 g (33.58 oz)
TOTAL DOUGH WEIGHT				4846 g (170.93 oz)

Procedures

		Yeast Sponge	Final Dough
Mixing: Low speed		6 min	4 min
Beater: Second speed			10 min
Dough temperature from mixer		(25° C / 77° F)	(25° C / 77° F)
Rest period for yeast sponge		4 h @ 24° C / 75.2° F or 16 h @ +4° C / 39.2° F	
First fermentation			3 h 50 min
Degas (punch down) after			(2 h 30 min)
Division, rounding & rest period			40 min
Dough forming			
Second fermentation			2 h 30 min
Baking @ 200° C / 392° F			35 min
PRODUCTION TIME			7 h 50 min

*The original French text included a small amount of egg in the *levain*. However, while it might be safe to add pasteurized egg to a sponge that is to be **refrigerated**, the danger of Salmonella contamination makes it inadvisable to use raw egg for this purpose, and neither raw nor pasteurized egg should be used in a sponge that will be kept at ambient temperature for any length of time. It is certainly preferable from a food safety point of view to simply add all the egg to the final dough, as has been done in this formula.

---

method used is a sponge and dough. Since the sponge has a high yeast content, it rises rather rapidly.

Before the sponge has finished its fermentation, the baker begins mixing all of the other dough ingredients, with the exception of the orange peel. The sponge is added to the dough mixture halfway through the mixing cycle. The mixed dough is then set to ferment at ambient temperature.

Exhibit 14–15 Pompe des Rois—Three Kings' Feast Cake by Sponge and Dough Method

Formula

	Weight		Yeast Sponge	Final Dough
Strong flour	2 kg (70.55 oz)		480 g (16.93 oz) (24%)	1.52 kg (53.62 oz)
Salt	32 g (1.13 oz)	1.6%		32 g (1.13 oz)
Yeast	80 g (2.82 oz)	4%	12 g (0.42 oz)	68 g (2.40 oz)
Sugar	600 g (21.16 oz)	30%		600 g (21.16 oz)
Butter	400 g (14.11 oz)	20%		400 g (14.11 oz)
Eggs*	600 g (21.16 oz)	30%		600 g (21.16 oz)
Milk	320 g (11.29 oz)	16%	320 g (11.29 oz)	
Raisins	400 g (14.11 oz)	20%		400 g (14.11 oz)
Candied fruit	400 g (14.11 oz)	20%		400 g (14.11 oz)
Lemon essence	(4 drops)			(4 drops)
Ascorbic acid	(30 mg)	(15 ppm)		(30 mg)
Pearl sugar (decorative, not added to dough)	(80 g [2.82 oz])	4%		
YEAST SPONGE			812 g (28.64 oz)	812 g (28.64 oz)
TOTAL DOUGH WEIGHT				4832 g (170.44 oz)

Procedures

			Yeast Sponge	Final Dough
Mixing: Low speed			8 min	4 min
Beater: Second speed				10 min
Dough temperature from mixer			(25° C / 77° F)	(26° C / 78.8° F)
Rest period for yeast sponge			3 h @ 24° C / 75.2° F *or* 15 h @ +4° C / 39.2° F	
First fermentation				4 h 30 min
Degas (punch down) after				(3 h 30 min)
Division, rounding & rest period				40 min
Dough forming				
Second fermentation				2 h 30 min
Baking @ 210° C / 410° F				35 min
PRODUCTION TIME				8 h 40 min

*As noted previously, the original French text included a small amount of egg in the *levain*. (See Exhibit 14–14.)

After rising, the dough is divided, the individual dough pieces are lightly rounded, and after a brief rest are rolled out into ovals about 30 cm long by 25 cm wide and 15 mm thick. Following this step, a small circle is traced in the center of the oval. Starting from the circle, use a dough blade or a razor to cut five slits in a star shape. These slits are enlarged to form eye-shaped gashes by pulling lightly on the outside edges

Exhibit 14–16 The "Gibassier" Hearth Cake of Provence by Sponge and Dough Method

Formula

	Weight		Yeast Sponge	Final Dough
Strong flour	2 kg (70.55 oz)	100%	400 g (14.11 oz)	1.6 kg (56.44 oz)
Salt	30 g (1.05 oz)	1.5%		30 g (1.06 oz)
Yeast	100 g (3.53 oz)	5%	50 g (1.76 oz)	50 g (1.76 oz)
Sugar	400 g (14.11 oz)	20%		400 g (14.11 oz)
Olive oil	300 g (10.58 oz)	15%		300 g (10.58 oz)
Margarine	300 g (10.58 oz)	15%		300 g (10.58 oz)
Eggs	500 g (17.64 oz)	25%		500 g (17.64 oz)
Milk	240 g (8.47 oz)	12%	240 g (8.47 oz)	
Orange flower water	100 g (3.53 oz)	5%		100 g (3.53 oz)
Orange peel	120 g (4.23 oz)	5%		120 g (4.23 oz)
Anise	40 g (1.41 oz)	2%		40 g (1.41 oz)
YEAST SPONGE			690 g (24.34 oz)	690 g (24.34 oz)
TOTAL DOUGH WEIGHT				4.13 kg (145.68 oz)

Procedures

Mixing: Low speed			6 min	4 min
Beater: Second speed				10 min
Dough temperature from mixer			(25° C / 77° F)	(25° C / 77° F)
Rest period for yeast sponge			45 min	
First fermentation				80 min
Degas (punch down) after				(50 min)
Division, rounding & rest period				35 min
Dough forming				
Second fermentation				70 min
Baking				25 min
PRODUCTION TIME				8 h 40 min

of the dough piece, and the hearth cake is set aside to proof.

Before baking, the cakes are brushed with egg wash and sprinkled with sugar crystals mixed with the anise noted in the formula. After baking, the gibassier is arranged and presented on wicker racks or screens.

BRIOCHE-TYPE HEARTH CAKES FROM OTHER LANDS

It is enlightening to follow a discussion of the principal types of French regional brioches with a brief look at a few yeast-raised, egg-enriched sweet dough hearth cakes made in Latin America and Italy, such as the Argentine *pan dulce* or sweet bread, a Brazilian brioche, and the Italian *panettone*. It should be noted that many other products of this type exist in Europe and elsewhere in the world, but that inquiry would be much too involved to pursue at the present time.

The *Panettone*

The *panettone* is a very rich and highly scented type of hearth cake made in Italy. They are little known in France, and to be quite frank, are not greatly esteemed there, even though they are generally of excellent quality. This lack of acceptance is doubtless because they seem to many Frenchmen (and women) to be too strongly scented and flavored. It is unfortunate that excessive use of orange flower water, orange essence, and vanilla seem to overpower the delicate flavor notes from alcoholic fermentation and the use of eggs and butter in the formula, but these practices are suited to Italian taste.

One of the distinguishing characteristics of the authentic *panettone* is that it is always made from a naturally leavened sponge, whether produced at the artisan level or on an industrial scale. This method of production allows the baker to produce a very rich product of excellent quality, which may be kept for weeks without any risk of its becoming moldy or losing its aroma and freshness, both of which become very evident whenever the cake is warmed slightly before being served.

For this reason the *panettone*, which is eaten primarily during the year-end holidays and on New Year's Day, can be made well in advance. It is widely exported from Italy to the large cities of both North and South America where there are sizable populations of ethnic Italian origin, such as New York or Buenos Aires.[7]

Other products that have formulations similar to the panettone are also widely made in Italy. The most common are

- the Pascal Dove, a brioche made in the form of a dove and prepared for the Easter holidays;
- the *pandoro*, which is made throughout the year and contains no preserved fruit, but does have a small amount of baker's yeast added to the dough;
- sweet dough croissants and *pandorinos* (small brioches), which are served at breakfast (the formulas for these products also contain a little baker's yeast).

These products are time consuming and difficult to prepare, but in spite of that are often made on an industrial level. In the case of *panettone* production, the greatest care should be given to the selection of the highest quality ingredients and to the culture of the naturally fermented sponges. Special effort should be directed to maintaining cultures that are very active and not overly acidic.

In standard production practice, these sponges are usually rebuilt or refreshed every 4 hours on a basis of 10 parts of mother sponge, 7.5 parts of water, and 15 parts of flour, or more simply, 4 parts of mother sponge, 3 parts of water, and 6 parts of flour. When the mother sponge is to be kept more than 6 hours, the most foolproof and practical method is simply to protect it from the drying effects of air and to place it under refrigeration at temperatures ranging from 8 to 10°C. Under these conditions it may be kept undamaged for 24 to 60 hours.[8]

Making up the leavening culture or *levain* from the mother sponge generally involves the preparation of three successive naturally fermented sponges (*levains*), each of which undergoes a 4-hour fermentation at 26–27°C. This process is further explained in Exhibit 14–17, and is similar to the method referred

---

Exhibit 14–17 Method of Building Naturally Fermented Sponges (Levains) for Panettone

Ingredient	1st sponge	2nd sponge	3rd sponge
Chef or mother dough	200 g (7.05 oz)	600 g (21.16 oz)	1.9 kg (67.0 oz)
Water	140 g (4.94 oz)	440 g (15.52 oz)	1.37 kg (48.32 oz)
Flour	260 g (9.17 oz)	860 g (30.34 oz)	2.73 kg (96.3 oz)
Total Weight	600 g (21.16 oz)	1.9 kg (67.0 oz)	6 kg (211.64 oz)
Fermentation time	4 h	4 h	4 h

to as *panification sur trois levains* (bread production from three natural sponges), based on an initial naturally fermented culture. Actual *panettone* production is also characterized by preparation of the dough in two or three stages, each of which includes staged addition of dough ingredients, especially sugar, and a suitable amount of mixing.

As previously noted, raw materials must be selected with great care. The baker should choose a strong flour with a protein content equal to or greater than 12%.[9] The eggs and butter should be of good quality and in excellent condition. For the most part only the egg yolks are used for this product. The use of egg whites is excluded, since they would have a drying effect on the *panettone* and adversely affect shelf life.

This industrial formula (Exhibit 14–18) also includes monoglycerides, which are used in the preparation of an emulsion (Exhibit 14–19) made from part

Exhibit 14–18 Formula Ingredients for Panettone (Industrial Type)

Formula

	Weight		Natural Sponge	Final Dough
Strong flour	29 kg (1022.93 oz)		4 kg (141.09 oz)	25 kg (881.83 oz)
Salt	290 g (10.23 oz)	1%		290 g (10.23 oz)
Sugar	8.12 kg (286.42 oz)	28%		8.12 kg (286.42 oz)
Honey	435 g (15.34 oz)	1.5%		435 g (15.34 oz)
Eggs (yolks only)	4.35 kg (153.44 oz)	15%		4.35 kg (153.44 oz)
Butter	7.83 kg (276.19 oz)	27%		7.83 kg (276.19 oz)
Monoglycerides	175 g (6.17 oz)	0.6%		175 g (6.17 oz)
Raisins	8.7 kg (306.88 oz)	30%		8.7 kg (306.88 oz)
Candied fruit	4.35 kg (153.44 oz)	15%		4.35 kg (153.44 oz)
Orange essence	(58 drops)			(58 drops)
Orange flower water	1.45 kg (51.15 oz)	5%		1.45 kg (51.15 oz)
Vanilla	29 g (1.02 oz)	0.1%		29 g (1.02 oz)
Water	8.7 kg (306.88 oz)	30%*	2 kg (70.55 oz)	6.7 kg (236.33 oz)
NATURAL SPONGE			6 kg (211.64 oz)	6 kg (211.64 oz)
TOTAL WEIGHT				73.429 kg (2590.08 oz)

Note: 24 parts of sponge (*levain*) is used for 100 parts of flour, or 24% of the flour noted for the final dough is used in the sponge.

Exhibit 14–19 Preparation of Emulsion for Panettone Production (Industrial Type)

Ingredient	Weight
Monoglyceride premix (175 g monoglycerides + 350 g of water @ 80° C / 176° F)	525 g (18.52 oz)
Egg yolks	2 kg (70.55 oz)
Butter	3 kg (105.82 oz)
Sugar	1.5 kg (52.91 oz)
TOTAL WEIGHT	7.025 kg (247.79 oz)

Dough Mixing Stages and Procedures

Ingredient	1st Mixing Stage	2nd Mixing Stage
Prefermented sponge	6 kg (211.64 oz)	
Flour	16 kg (564.37 oz)	9 kg (317.46 oz)
Salt		290 g (10.23 oz)
Sugar	3 kg (105.82 oz)	3.62 kg (127.69 oz)
(Sugar emulsion)	(1.5 kg [52.91 oz])	
Honey	435 g (15.34 oz)	
Eggs		2.35 kg (82.89 oz)
(Egg emulsion)	(2 kg [70.55 oz])	
Butter		4.83 kg (170.37 oz)
(Butter emulsion)	(3 kg [105.82 oz])	
(Monoglyceride emulsion)	(175 g [6.17 oz])	
Raisins		8.7 kg (306.88 oz)
Candied fruit		4.35 kg (153.44 oz)
Orange flower water		1.45 kg (51.15 oz)
Vanilla		29 g (1.02 oz)
Orange essence		(58 drops)
(Water emulsion)	(350 g [12.35 oz])	
Water	6.35 kg (223.99 oz)	
Dough weight (first mixing stage)	38.81 kg (1368.96 oz)	38.81 kg (1368.96 oz)
TOTAL DOUGH WEIGHT		73.429 kg (2590.09 oz)
Mixing @ first speed (90 rpm)	3 min	3 min
Spiral mixer @ 2nd speed (200 rpm)	8 min	8 min
Dough temperature	(25° C / 77° F)	(25° C / 77° F)
1st fermentation 1st dough	10–12 h	
1st fermentation 2nd dough		40 min
Division–rounding–rest		50 min
Rounding and depositing		
2nd fermentation @ 27° C / 80.6° F		10–11 h
Baking @ 210° C / 410° F		30–40 min
PRODUCTION TIME		13 h 20 min

of the egg yolks, sugar, butter, and a little water. This emulsion assists in the formation of the gluten film during mixing, improves the final volume of the product, and helps to preserve the freshness of the *panettone* under storage conditions.

To prepare the emulsion: mix together the noted ingredients until a homogeneous mass is obtained, comparable to a mayonnaise. The emulsion is added to other main ingredients in the first mixing stage.

The incorporation of raisins and candied fruits into the dough is done at low speed at the end of the second mixing stage. To facilitate mixing, hold back a small amount of softened butter to lubricate the dough and to avoid damaging the fruit.

The makeup of *panettone* includes two separate rounding stages: the first after dividing, and the other after the *détente* or rest period. The rounded dough ball is then deposited into a paper mold where it undergoes the second fermentation, at a temperature of 27°C / 80.6°F. Just before placing the dough into the oven, the baker cuts a cross shape into the upper dome surface of the *panettone*, and the resulting four corners become slightly raised. For a 500 gram cake, baking lasts approximately 40 minutes at 200 to 210°C / 392 to 410°F.

Cooling is usually done with the *panettone* suspended upside down, hanging from three steel pins inserted into the base. This practice avoids having the cake shrink or fall during cooling, and it also results in a slight elongation, which has the effect of increasing the height of the product. The cooled *panettones* are then sprayed with a fine mist of a special alcohol vapor mixture to sterilize their exterior crust walls, following which they are enclosed in an airtight wrapping. The alcohol spray is composed of a liter of 95% alcohol and 10 g of acetic acid, and used at a rate of 3 cc per *panettone*.

A *panettone* can be leavened with baker's yeast. However, this is not an authentic *panettone* and remains only a pale imitation of the real thing in terms of flavor and keeping qualities. If the baker chooses not to use a naturally leavened sponge, the sponge and dough system (Exhibit 14–20) is the least objectionable method.

It should be pointed out that in Italy there is a trade label for the authentic Italian *panettone*, which specifies the minimal level of required ingredients to be included in the dough and specifies that the

---

Exhibit 14–20 An "Imitation" Panettone by Sponge and Dough Method

Formula

	Weight		Yeast Sponge	Final Dough
Strong flour	5 kg (176.37 oz)		1.25 kg (44.09 oz) (25%)	3.75 kg (132.28 oz)
Salt	50 g (1.76 oz)	1%		50 g (1.76 oz)
Yeast	200 g (7.05 oz)	4%	25 g (0.88 oz)	175 g (6.17 oz)
Sugar	1.25 kg (44.09 oz)	25%		1.25 kg (44.09 oz)
Butter	1.25 kg (44.09 oz)	25%		1.25 kg (44.09 oz)
Whole milk powder	200 g (7.05 oz)	4%		200 g (7.05 oz)
Whole eggs*	1.5 kg (52.91 oz)	30%		1.5 kg (52.91 oz)
Water	800 g (28.22 oz)	16%	700 g (24.69 oz)	100 g (3.53 oz)
Orange flower water	250 g (8.82 oz)	5%		250 g (8.82 oz)
Orange essence	(10 drops)		—	(10 drops)
Vanilla	5 g (0.18 oz)	0.01%		5 g (0.18 oz)
Raisins	1.5 kg (52.91 oz)	30%		1.5 kg (52.91 oz)
Candied fruit	500 g (17.64 oz)	10%		500 g (17.64 oz)
Ascorbic acid	(200 mg)	(20 ppm)		(200 mg)
YEAST SPONGE			1.975 kg (69.66 oz)	1.975 kg (69.66 oz)
TOTAL DOUGH WEIGHT				12.505 kg (441.09 oz)

Procedures

Mixing: Low speed		6 min	4 min
Beater: Second speed			10 min
Dough temperature from mixer		(25° C / 77° F)	(26° C / ° F)
Rest period for yeast sponge		(4 h @ 25° C / 77° F or 15 h @ +4° C / 39.2° F	
First fermentation			100 min
Degas (punch down) after			(50 min)
Division, rounding & rest period			40 min
Dough forming			
Second fermentation			140 min
Baking (@ 210° C / 410° F)			40 min
PRODUCTION TIME			5 h 35 min

*Whole eggs are used in this formula, rather than yolks alone. Professor Calvel notes in the text that this will produce a drier crumb with a reduced shelf life.

term *panettone* cannot be used for such a bakery product unless it has been leavened with a naturally fermented sponge.

Brazilian Brioche

The formula in Exhibit 14–21 was developed during a training course for professors of bakery science and bakery technologists that I gave in November of 1982 at Fortaleza, in the northeast of Brazil. The dough was flavored with the fruit and flowers of Brazilian oranges.

This particular dough may be fashioned into a number of shapes: crowns, rounded or elongated loaves, or rounded and placed in molds after division. It may also be used for smaller dough units, to be formed into rounded or shuttle-shaped (i.e., boat shaped) rolls. Whatever their form, the dough pieces are to be brushed with egg wash and may also be sprinkled with pearl sugar before being placed into the oven.

Argentine Sweet Bread (Le Pan Dulce Argentin)

As its name indicates, this sweet bread is a bakery product similar to a brioche in that the dough is contains a high percentage of eggs (Exhibit 14–22). All of the fruits and nuts in this cake are produced in Argentina, including raisins, candied fruit, nuts, and almonds. To have a light crumb structure, this product requires a strong flour and is improved by being leavened with a prefermented culture. This may quite simply be just a well-fermented ordinary bread dough.

Exhibit 14–21 A Brazilian Brioche by the Sponge & Dough Method

Formula

	Weight		Yeast Sponge	Final Dough
Strong flour	2 kg (70.55 oz)	100%	500 g (25%) (17.64 oz)	1.5 kg (52.9 oz)
Salt	36 g (1.27 oz)	1.8%		36 g (1.27 oz)
Yeast	140 g (4.94 oz)	7%	10 g (0.35 oz)	130 g (4.59 oz)
Sugar	500 g (17.64 oz)	25%		500 g (17.64 oz)
Eggs	600 g (21.16 oz)	30%		600 g (21.16 oz)
Milk	350 g (12.35 oz)	17.5%	350 g (12.35 oz)	
Butter	300 g (10.58 oz)	15%		300 g (10.58 oz)
Margarine	200 g (7.05 oz)	10%		200 g (7.05 oz)
Orange flower water	80 g (2.82 oz)	4%		80 g (2.82 oz)
Orange flavoring	16 g	0.8%		16 g
Ascorbic acid	(100 mg)	(40 ppm)		(100 mg)
YEAST SPONGE			860 g (30.34 oz)	860 g (30.34 oz)
TOTAL DOUGH WEIGHT				4222 g

Procedures

		Yeast Sponge	Final Dough
Mixing: Low speed		6 min	4 min
Beater: Second speed			10 min
Dough temperature from mixer		(26° C / 78.8° F)	(26° C / 78.8° F)
Rest period for yeast sponge		(4 h @ 27° C / 80.6° F or 15 h @ +4° C / 39.2° F	
First fermentation			220 min
Degas (punch down) after			(140 min)
Division, rounding & rest period			30 min
Dough forming			
Second fermentation			140 min
Baking (@ 210° C / 410° F)			35 min
PRODUCTION TIME			7 h 20 min

Exhibit 14–22 Argentine Sweet Bread from Prefermented Dough

Formula

	Weight		Prefermented Dough	Final Dough
Strong flour	2 kg (70.55 oz)	100%	400 g (14.11 oz)	1.6 kg (56.44 oz)
Water	240 g (8.47 oz)	12%	240 g (8.47 oz)	
Salt	40 g (1.41 oz)	2%	8 g (0.28 oz)	32 g (1.13 oz)
Yeast	100 g (3.52 oz)	5%	10 g (0.35 oz)	90 g (3.17 oz)
Butter	400 g (14.11 oz)	20%		400 g (14.11 oz)
Sugar	440 g (15.52 oz)	22%		440 g (15.52 oz)
Eggs	700 g (24.69 oz)	35%		700 g (24.69 oz)
Milk	100 g (3.53 oz)	5%		100 g (3.53 oz)
Orange flower water	80 g (2.82 oz)	4%		80 g (2.82 oz)
Almonds	100 g (3.53 oz)	5%		100 g (3.53 oz)
Nuts	100 g (3.53 oz)	5%		100 g (3.53 oz)
Raisins	500 g (17.64 oz)	25%		500 g (17.64 oz)
Candied fruit	400 g (14.11 oz)	20%		400 g (14.11 oz)
Ascorbic acid	(75 mg)	(30 ppm)		(75 mg)
YEAST SPONGE			658 g (23.21 oz)	658 g (23.21 oz)
TOTAL DOUGH WEIGHT				5.2 kg (183.42 oz)

Procedures

Mixing: Low speed	3 min	4 min
Beater: Second speed	10 min*	10 min
Dough temperature from mixer	(25° C /77° F)	(26° C / 78.8° F)
Rest period for yeast sponge	(4 h @ 25° C / 77° F or 15 h @ +4° C / 39.2° F	
First fermentation		160 min
Degas (punch down) after		(100 min)
Division, rounding & rest period		35 min
Dough forming		
Second fermentation		140 min
Baking (@ 210° C / 410° F)		40 min
PRODUCTION TIME		6 h 30 min

*The original French text mistakenly gave 100 min as mixing time.

NOTES

1. It is very possible that all of these began as sourdough recipes.
2. The order of the recipes in this chapter is slightly different from the French text. Since Professor Calvel has often expressed to James MacGuire his great fondness for the kugelhopf, it is given pride of place in this edition.
3. Many French recipes that call for bacon pieces indicate that the bacon should be "blanched" by bringing it to a boil in cold water before browning.
4. In contrast to standard American practice, the French generally use a slightly fermented cream (crème fraîche) for whipping. This provides a richer and more complex flavor profile, as well as a more acidic pH, which has greater resistance to bacterial growth and is conducive to longer shelf life.
6. This custom still survives in Italy at the present time.
7. The popularity of the *panettone* has also spread to many other Latin American countries with large populations of European origin, including Colombia, Peru, Venezuela, Paraguay, and Chile.
8. Temperatures lower than 8 to 10°C would tend to damage the culture.
9. The original French text specified a type 55 wheat flour with a W value of around 280, with a P/L of 0.8. In actual practice, many North American bread flours would be acceptable. A Technical Bulletin (1992, No. 8) available from the American Institute of Baking discusses at some length the production of industrial-type *panettone* with North American flours.

Part VI

Nutritional Value of Bread, Bread and Gastronomy, Bread and the Consumer

Qualities of Bread

Chapter 15

- The nutritional value of bread.
- Caloric content and bioavailability.
- The progressive decline of bread consumption in France.
- Bread made from stone-ground flour.
- Bread and gastronomy.
- Comparing bread with other foods.

THE NUTRITIONAL VALUE OF BREAD

Bread has a high energy value, just as does the flour from which it is made.[1] It is also a carrier for some other non-caloric nutritional elements that are completely indispensable to proper nutrition. These nutritional elements vary according to the purity and the extraction rate of the flour.

The energy-providing components of bread and flour are

- carbohydrates (starches, sugars), which provide around 4 calories per gram
- lipids (various types of fats), which supply around 9 calories per gram
- proteins (also found in meats, milk, legumes, and cereals), which provide around 4 calories per gram.

The other principal nutrients provided by flour and bread include some fatty acids and amino acids, a number of minerals, and vitamins A, B_1, B_2, C, D, E, niacin, and K.

Table 15–1 notes the principal nutrients furnished by three types of bread made from flours milled at different extraction rates. All of the figures are for 100 g samples of bread at 30% to 34% humidity.

Traditional white French bread is made from type 55 wheat flour, milled at an extraction rate of 75 to 78%.

The energy content provided by the average per capita consumption (in France) of 145 g (5.11 oz) of bread per day amounts to a total of 375 calories, or 17% of the average daily nutritional requirement, estimated at 2200 calories per day. There are slight differences in caloric or energy content between the three types of bread noted in the table. On a practical level, these differences are due as much to variations in the water content of the types of bread as to differences that derive from the rate of extraction.

However, it should be stressed that bread made from white flour has the best coefficient of nutritional utilization. It is true that bread made from white flour has a lower mineral and vitamin content compared with breads made from darker flours (extracted at an 85% rate), and especially with whole-wheat flours (extracted at a 98% rate), and it is not uncommon for those in some nutrition-related fields to scorn white flour for this reason.[2]

On the other hand, what criticisms might be made of these other flours? What about real whole-wheat bread made from an authentic whole-wheat flour? It has a high moisture content, rises badly, has a crumb

Table 15–1 Principal Nutrients Furnished by Three Types of Bread Made from Flours Milled at Different Extraction Rates

Rate of Extraction	Calories / 100 g	Principal Energy Contributors in Grams			Mineral Content in Milligrams			Vitamin Content in Grams		
		Carbs	Lipids	Proteins	Phos	Cal.	Iron	B_1	B_2	PP
98% (whole)	245	7	1	52	180	57	2.5	0.22	0.25	4.30
85% (high)	252	7	1	53	120	30	1.7	0.15	0.12	0.65
78% (normal)	260	7	1	54	91	20	1	0.06	0.06	0.43

that is too tightly structured, and has a strongly pronounced bitter bran taste—in short, the quality is inferior.

It is true that the high cellulose (bran) content makes it an excellent remedy for constipation. However, even if the wheat is very carefully cleaned, the inclusion of the entire bran envelope in flour entails the risk of increasing the ingestion of pesticides and fungicides that are used in wheat cultivation and storage. Furthermore, whole-wheat flour's high phytic acid level causes dietary calcium loss,[3] and for some individuals it has a tendency to irritate the mucous membranes of the intestines, thereby lowering the nutritional utilization coefficient of the rest of the foods in the diet.[4]

Bread made from dark wheat flour milled at an 85% extraction rate has two times *less* phytic acid than real whole-wheat flour, but two times *more* phytic acid than white bread, which is a not a minor disadvantage.[5] Although dark wheat flour does not cause any digestive problems in particular, it is higher in pesticide residues than white flour. Finally, dark wheat flour still has the problem of a notable bitter bran taste, which degrades the quality and is atypical of good bread.

It should be added that flours with high bran levels, including both whole-grain flours and high-extraction dark flours, produce doughs that are physically weak, with a marked tendency toward stickiness and sliminess, a more porous structure, poor fermenta-

tion gas retention, and more limited tolerance. All other conditions being equal, breads from these high-extraction flours have less volume, have a more tightly structured crumb, and—at least in my estimation—are generally less pleasant to the taste.

In 1977 I carried out a series of breadmaking tests, without the use of any additives, on three grades of flour *milled from the same wheat*. I have repeated the same tests on other occasions. The three flour grades—a type 55 breadmaking flour, a type 80 dark wheat flour, and a type 150 whole-wheat flour—produced breads with significant differences in volume. The type 80 dark wheat flour produced loaves that were 19% smaller in volume than those made from the type 55 bread flour. The volume of the loaves made from type 150 whole-wheat flour was a little more than 50% lower than that of the type 55 bread.

Similar results in actual production of doughs and breads make breadmaking from these types of high-extraction flours very difficult and time-consuming. In order to simplify the process and to obtain larger loaf volume, bakers often find it necessary to increase additive use. Furthermore, since bread from these flours has a higher moisture content, shelf life may be more limited, and spoilage inhibitors must often be used in addition to those additives used in production.

CALORIC CONTENT AND BIOAVAILABILITY

In any consideration of bread consumption and the physiological value of including all or part of the bran envelope in bread, it is entirely reasonable to ask whether humans are able to take advantage of the higher levels of vitamins and minerals in the bran coating.

Toward the end of the 19th century, chemist Aimé Girard, who was a professor at the Institute of Agronomy, showed that bran passed through the digestive system remains in much the same state as when it was ingested and concluded that it did not seem to be worthwhile to consider it as a nutritional factor.

Certainly, if one accepts at face value the chemical analyses carried out on this subject and summarized in the Table 15–2, it seems that white bread is less nutritional—unless one takes into account the fact that much of the fiber is indigestible. The figures cited are given in terms of relative nutritional value and are based on a study by Causeret. They were first presented in a paper given by Professor H. Gounelle de Pontanel and Dr. Pradini-Jarre at the CNERA colloquium on bread in 1977. The figures represent the principal vitamins and minerals of the breads examined in the study.

As far as proteins, fats, and carbohydrates are concerned, there are few notable differences. The amount of protein and fat is practically identical for white flour, high-extraction dark flour, and whole-wheat flour, while the carbohydrate content is greater for breads made from type 55 bread flour, as evidenced in the table.

After determining the values for vitamins and minerals that are attributed to the bran coating, Professor de Pontanel and his collaborator noted that these values varied from a relative level of 1 to 3 within the wheat kernel—a considerable variation. The researchers asked as well—without being totally convinced of it—if that was sufficient reason to attach greater importance to whole-wheat bread.

They also gave the amount of phytic acid supplied by the bran envelopes and noted its deleterious effect on the absorption of minerals such as calcium, mag-

Table 15–2 Chemical Analysis of Principal Vitamins and Minerals

Nutrient	Average amount from the normal daily food intake for moderately active adult (mg/day)	Average amount supplied daily by 300 g of bread		
		Whole-wheat flour 98% extraction	(Dark) wheat flour 82–85% extraction	Type 55 bread flour 77–78% extraction
Vitamin B_1	1.5 mg	50%	35%	15%
Vitamin B_2	1.8 mg	20%	12%	7%
Niacinamide	15 mg	80%	20%	13%
Vitamin E	10–15 mg	30%	25%	20%
Magnesium	350 mg	80%	30%	15%
Iron	15 mg	50%	20–25%	15%
Copper	1.5 mg	90%	40%	25%

nesium, and iron. On one hand, they did not overemphasize the harmful effects of phytic acid, since (1) the body seems to become somewhat accustomed to it over time, and (2) a prolonged dough fermentation seems to reduce its toxic factors. On the other hand, however, they did not rule out the risk of some type of toxicity. They also believe that research studies are needed to determine the maximum level of bran particles that can be present in flour without causing limitations on the proper biological use of minerals. They also add:

> "Our knowledge regarding the tolerance of different classes of consumers for the indigestible fibrous components of breads which are more or less whole-grain, and concerning the actual effects of this added fiber on nutrition, must be increased...."

In passing, they also mention the beneficial effects of crude fiber on the proper functioning of the digestive system. The beneficial effects of wheat bran are especially suitable to those individuals with "lazy" digestive systems—fortunately, they are not in the majority.

Finally, the researchers stress that it would be an error to judge bread on the basis of its distinctive nutritional balance. Bread is a food especially rich in carbohydrates. These carbohydrates are largely made up of complex starches, which are absorbed slowly by the body. They do not contribute to the rapid rise in blood sugar characteristic of hypoglycemia, and furthermore they are very low in fat. It is these characteristics that make bread an especially desirable food from a nutritional perspective.

I should add that as a consumer I am greatly skeptical regarding the nutritional advantages of the vitamins and minerals provided by the presence of wheat bran in flour, as presented to us in the tabulated results of various analyses of bread. We have known since Aimé Girard took the trouble to prove it to us that the human digestive system is incapable of digesting wheat bran. Even though many of the cellulose cell walls in bran are destroyed in milling, what are the benefits of ingesting those vitamins and mineral salts that are for the most part are found unchanged after excretion?

For the past decade, a great deal of publicity has promoted the advantages of dark wheat breads and whole-wheat breads, based on the opinions of certain medical and pseudo-medical groups that praise the virtues of fiber consumption. It seems to me that it is completely unreasonable to advance these opinions and to attempt to induce others to share them. Even today, none of these opinions on the value of fiber seem to have a very solid foundation, except in the case of "lazy" intestine at the clinical level. On the other hand, it has certainly been proven that for a sizable part of the population—notably children up to 10 or 12 years of age—the consumption of dark wheat breads, and especially whole-wheat breads, is to be highly discouraged.

Professor Lestradet is a well-known pediatrician whose authority is beyond question. In a recent presentation on the value of bread in child nutrition, he gave his opinion on the type of bread that should be given to them:

> "Personally, on the basis of both taste and experience, I believe that we must give normal white bread to children, not whole-wheat bread. Some of our colleagues who are physicians for adults sometimes prefer whole-grain bread for their patients, but it is preferable to give them traditional white bread, especially country-style bread, which often has superior taste. Whole-wheat bread contains an excessive amount of bran. It also has the principal disadvantage of serving as a carrier for phytic acid, which retains calcium, iron, and magnesium in the feces. This is definitely not the desired effect."

In his later discussion about the food value of bread, Professor Lastradet added:

> "In general, ordinary bread includes a remarkable number of fibers which slightly retard the absorption of sugars. Viewed in this respect, it does not seem that whole-wheat bread has any particular advantages over common white bread."

He further concluded that "bread should occupy a special place in diets of children."

Because of the interest that consumers have in our daily bread, I would like to add some short extracts from presentations given by medical doctors and professors within the field of the health sciences who took part either in the 1977 Bread Colloquium or in the 1980 Colloquium on the Value of Bread in the Diet. I have also added a few other related commentaries.

I observed everything that was written and said in the course of these proceedings with the greatest interest and the highest degree of respect. However, it seemed to me that generally the participants spoke about bread in terms and a manner that were much too coldly clinical. This conclusion doubtless comes from my feeling as a baker who delights in making bread that is both delicious to eat and pleasant to behold, and from the fact that as a consumer I value these properties even more. Considering these inclinations, I especially valued the presentations given by the following experts:

- Professor Dupin stated about bread that "...it is a food which has lost none of its nutritional attractiveness. We must make an effort to restore its rightful importance by a consumer education initiative that is objectively and scientifically based."
- Professor Lestradet favored the advancement of the notion that for children "...tea or afternoon snack time should be bread and jam or bread and chocolate and not a fat-filled cake which tends to upset the balance of good nutrition."
- Professor Vague advised that "a ration of from 300 to 350 g of bread for a man (which represents from 650 to 700 calories) or from 200 to 250 g for a woman (from 430 to 450 calories) is a perfectly acceptable daily dietary intake." He especially prefers the slowly absorbed complex carbohydrates from bread to the more rapidly absorbed simple carbohydrates from sugars and sugar-based products.
- Doctor Plat stressed the importance of developing a program to inform and educate young consumers about good nutrition and to encourage them to eat more bread, less sugar, less meat, and fewer high-fat foods.
- Doctor Bour advised caution in the consumption of cellulose-type fibers, such as those found in bran and contained in whole-wheat bread and bran bread—except in the case of those suffering from constipation.
- Doctor Plat underlined the risks that may occur from the consumption of short-fermentation breads made from dark, high-extraction flours and whole-grain flours because of the possible intake of excessive amounts of phytic acid.

While still holding to the idea that consumption of a food like bread is not increased by legislating it, and that consumers will eat more bread when it is appetizing and good tasting, I would like to add that several presentations left me a little skeptical, or even a little bit unsettled:

- Professor Vigne noted that bread plays an important role in providing various vitamins and minerals to the body, and that the tendency of the consumer to continually look for whiter and whiter bread may deprive him or her of part of these nutritional elements. I do not believe that consumers search for "whiter and whiter bread," but that they are simply looking for good bread, the authentic quality product that we as bakers

should offer to them. How much would we (and they) consume if it were offered as dark wheat bread or as whole-grain bread? Is it in the interest of the consumer to consume more good bread, or to consume less in the form of dark wheat bread—or even to give up eating it altogether?
- Doctor Plat prudently suggested an increase in the rate of extraction from 75% (which is in actual practice about 77%) to 82%–85%, and to use type 80 flour (or even type 110 flour) in the place of type 55 bread flour.

These proposals were inspired by good, if debatable, intentions, but they could have unfortunate results in the end. Just as Professor de Pontanel did during the 1977 Bread Colloquium, and before abandoning our traditional white bread in favor of an intermediate or whole-wheat bread, we should ask ourselves if we can answer a number of questions affirmatively:

- Would the use of flour with an extraction rate greater than 75% (actually 77% to 78%) to make bread not result in an increase in the level of pesticide residues?
- Would a type of bread much higher in indigestible fibers be suitable for all consumers, and in particular for children?[6]
- Would this type of bread be acceptable to French consumers?[7]

THE PROGRESSIVE DECLINE OF BREAD CONSUMPTION IN FRANCE

The consumption of bread in France has experienced a constant decline from the end of the 19th century until the present. The daily and per capita consumption has undergone the following changes:

Time period	Amount consumed per person per day
During the 18th and 19th centuries	750 g to 1 kg (26.5 to 35.27 oz)
from 1890 to 1914	600 g (21.16 oz)
from 1920 to 1929	600 g (21.16 oz)
from 1929 to 1935	450 g (15.87 oz)
from 1937 to 1939	350 g (12.35 oz)
from 1950 to 1951	300 g (10.58 oz)
from 1960 to 1962	260 g (9.17 oz)
from 1970 to 1972	195 g (6.88 oz)
from 1980 to 1982	160 g (5.64 oz)
1985	145 g (5.11 oz)

In parallel with the decrease in bread consumption, it should be pointed out that the average daily caloric consumption per capita has also diminished a great deal. It has declined from:

Before 1914	3500 calories
from 1930 to 1935	3000 calories
from 1937 to 1939	2500 calories
since 1960–1961	2400 calories
today, approximately	2200 calories

The amount of daily energy intake due to bread (100 g of bread contributes about 260 calories) from 1900 to 1935 was equivalent to about 43%. From about 1937 the daily per capita energy intake from bread declined to 36%. Today it hardly even equals 20%.

The Causes of the Decline in Bread Consumption

There are a number of reasons for the decline in bread consumption, and for the most part they are linked to mechanization of the factors of production and the consequent decline in physical effort, which emerged rapidly during the first half of the 20th century. From the 1930s on—with the exception of the World War II period—this decline in consumption has also been stimulated by the improvement in general quality of life resulting from mechanical progress. The quality and form of bread, as well as the periodic anti-bread campaigns that occurred from the 1930s through the 1950s, have also been factors in this consumption decline.[8]

There has been a lowering of the level of physical activity in all its forms because of mechanization in industry, agriculture, and domestic tasks, down to the level of the abandonment of walking and the habitual use of elevators in favor of climbing stairs. The change extends to methods of heating buildings, both at home and in the workplace, which has also very markedly reduced the need for food energy.

Because of the general increase in the quality of life, the consumer has many more opportunities to diversify his or her normal diet and to replace bread with more costly foods such as meats, eggs, fish, cheeses, vegetables, and fruits. Changes in the mode of life have caused the gradual abandonment of bread, which accompanied soups and those meats cooked in sauces, which have been replaced by grilled meats.

The quality and other characteristics of bread have also contributed to a decline in the consumption of bread during certain historical periods. This was especially true during the time of severe economic hardship that France experienced from 1940 until August 1947 because of World War II. This entire period was marked by the production under very difficult conditions of poor-quality bread from what were more or less whole-grain flours, or whole-grain flours mixed with high proportions of non-wheat flour substitutes.[9]

The use of these poor-quality ingredients for eight long years contributed greatly to discrediting bread and also seriously hampered proper professional training. Bakers were obliged to adopt emergency production methods and to train young apprentices under abnormal conditions. Even after the end of this trying time, the difficult period of readapting to sound technical methods continued to reduce bread quality somewhat.

The year 1958 marked the appearance of rapid mixing or intensified mixing technology, which was adopted in only four or five years by nearly all professional French bakers. This resulted first of all in the production of loaves that were much larger in volume, with a much whiter crumb and much greater appeal to both bakers and consumers.

Second, intensified mixing contributed to the general adoption of bean flour and ascorbic acid as bakery additives and to the use of more mechanical equipment for the weighing and shaping of dough pieces more rapidly. This resulted in loaves with even greater volume but decreased shelf life, and with an atypical and insipid, often disagreeable, taste.

Third, this practice of producing breads larger in volume for a given weight caused the consumer to buy his or her bread more on the basis of volume rather than of weight, thereby contributing to an automatic lowering of bread consumption.

This evolutionary change, including the development of excessive bread volume, adulteration and lessening of taste, shortening of shelf life, and the other factors already discussed above culminated in a spectacular lowering in bread consumption. This reached its lowest point during 1962–1968: a decline of nearly 4% in the Parisian region and nearly 3% on a national level. While the reduction in average bread consumption has continued, the rate of decline has become more moderate.

Beginning in the 1980s, a reaction began to take form within the ranks of professional bakers. This movement favored the production of a better type of bread, one that would better fulfill the desires of the consumer for a product with a creamy white crumb, a more distinctive flavor, better shelf life, and less volume. The movement spread gradually, and it was only in the autumn of 1983 that the potential importance of this trend to the production and consumption of bread was finally taken into account by the entire French baking industry. It should be added, however, that this realization has been transformed into action very slowly.

Yet another factor in the decline of bread consumption has been the vilification to which it has been subjected over the past 30 or 40 years. It has been said by some that bread today is made from flours that are inferior to those used in the past, that it is nutritionally deficient and indigestible, that it is made through the use of chemical additives, and that it causes excessive weight gain. We can certainly reply that these claims have no basis in fact and are rather the result of ignorance or deliberate slander.

From 1954 through 1960, a commission chaired by Professor Terroine, director of the Center for Research on Food and Nutrition, met periodically to consider these various problems and to respond to these accusations. This commission was made up of physiologists, chemists, medical doctors, milling and baking technicians, and wheat producers.

Since that time, other studies conducted by members of the medical profession with a special interest in nutrition have agreed with the conclusions of the commission and have brought to light the real nutritional value of bread. It has been shown to be a very healthful food, with the highest degree of nutritional value, and it would be desirable to increase the average per capita consumption for both children and adults to a level of 350 to 400 g per day.[10]

This interesting concept could be as beneficial for consumers as it is for bakery professionals, flour millers, and wheat growers, but cannot be (undertaken) unless the product that the baker offers is a true quality bread, both appetizing and tempting.

BREAD MADE FROM STONE-GROUND FLOUR

Certain groups praise stone-ground flour highly, presenting it as synonymous with tradition, authenticity, and quality, while trying to evoke an imaginary, far-distant past. In their case, this is simply exploitation of an imaginary but very fruitful "mother lode."

Following the first appearance of roller mills at the end of the last century, the French milling industry became concerned about quality problems. Consequently, a number of long-term, rigorous tests were conducted at the end of the 1880s on the flours produced by means of both roller and stone milling methods.

Beginning with identical lots of wheat, everything was compared, including breadmaking—and the flour produced with the roller mill won the competition. This flour produced the best and most attractive loaves, in an era when consumers knew how to appreciate good bread, and the average per capita bread consumption was 600 g per day.

What is there to say today, except that the results of the milling tests conducted in the 1880s are still valid? All other conditions being equal, the roller mill still produces high-quality breadmaking flours and breads that are generally superior to those that come from stone mills.

BREAD AND GASTRONOMY

What is the place of bread in the world of gastronomy? Before even trying to reply to that question, especially today, it is necessary to ask another question—that is, to what bread are we referring? As far as I am concerned, there can be only one answer. It can only be real bread—that is, bread made from the traditional mixture of flour, water, salt, and a fermenting agent, either baker's yeast or a naturally fermented sponge and produced according to the true nature of things and the rules of the baker's art.

Such a bread results from the following:

- the use of a good breadmaking flour that is made only from wheat;[11]
- the makeup of a dough by means of suitable mixing practices, and the production of a suitable gluten matrix without excessive and harmful oxidation;
- an adequate and well-regulated fermentation, scrupulously carried out in two stages;
- dough division and loaf forming practices that protect the dough from excessive physical stress;
- proper baking of formed loaves that are reasonably well proofed and that do not suffer from either insufficient or excessive development;
- crust color that has the regular golden yellow tint common to breads that have been properly baked and
- a crumb structure that is creamy white, with large and numerous irregularly-spaced cells.

This is the bread I tried to depict in the small poem I composed toward the end of the summer of 1987:

What is bread?

Bread is the wheat the farmer sows,
And the grain that awakens beneath the earth.
Bread is the pale shoot that rises,
And the blade that breaks forth in the Spring.
Bread is the fertile flower, and the head that forms,
And the wheat berry that swells and ripens.
Bread is the golden harvest in itself,
And the wheat gathered in, and the grain in the silo.
Bread is the grain in the mill,
And the wheat separated from the chaff, and the
  crushed grain.
Bread is the wheat transformed,
And the milled flour stream passing through the sieve.
When the pure flour has been extracted, and carried
  to the bakery,
Bread is the baker at his work, and the mixer in
  action.
It is the fully prepared dough, rich with fermented life.
Bread is good dough, kneaded just as it should be,
and just as lovingly fermented.
Bread is the dough, carefully divided and weighed,
And the living loaves, formed with gentle care.
Bread is well-proofed loaves that are finished
By a good and proper baking.
Bread is the sum of all these necessary things...
Wheat of good quality, millers with great skill,
And above all those dedicated, artful bakers
Who taste the bread they have so carefully created.
It is thus that bread, good bread, is made from pure
  wheat flour
That first part proceeds from Nature, and the rest from
  Man.
Bread is those who are consecrated to the craft,
All those who respect the nature of things, and the
  rules of the art.
Bread is all of that, and all that forms its history
And let us never doubt that it will also be its future.[12]

Although these basic principles are valid both for bread in general, and for French bread in particular, we must make distinction in terms of taste and eating quality between bread leavened with a natural sponge and that made with baker's yeast. Bread made from a naturally fermented sponge or "sourdough" has a light cream-colored crumb with grayish tones, a very definite and distinctive odor and taste, and a sharp acetic acid flavor provided by the complex organic acids which are derived from the products of fermentation. This bread has a wholesome, rustic flavor and aroma. It is pleasant to chew and has especially attractive eating qualities.

Bread leavened with baker's yeast has a golden crust, a creamy white crumb, and a pleasant and attractive combination of aroma and flavor. This complex of aroma and taste unites the combined scent of wheat flour with those of alcoholic fermentation and caramelization of sugars in the crust to give this type of bread highly agreeable eating properties and wonderful taste. As we have noted previously for both types of products, such results can only be attained by rigorously following the prescribed rules.

These two types of product did not originate very long ago in relation to the total history of bread. They materialized around the beginning of this century, and until the 1920s were known respectively by the names "French bread" and "Vienna bread." The first was always leavened with a natural sponge, generally without any added baker's yeast, although a small amount was sometimes used. The second was leavened with baker's yeast, used in the form of a prefermented culture called the "*poolish*" or Polish style sponge.

In rural areas especially, around the early 1920s, the practice of making bread with natural leavening during the cold season began to change. Bakers began to use the prefermented dough method, which included the use of a significant amount of baker's yeast. In the cities, about the same time, the straight dough method using baker's yeast replaced the prefermented Polish style sponge. Use of the "*poolish*" became somewhat rare, and the terms "French" bread and "Vienna" bread began to disappear.

During this period of mutation and evolution, dough division and forming were still manual operations, and numerous forms and types of bread remained in existence. These bread types and loaf forms had been inherited from the era when the guilds required each journeyman candidate for advancement to produce a new shape and style of loaf from a portion of dough. However, from the 1920s on, these different types of loaves were produced from yeast-leavened doughs, all of which were very similar in shape if not identical.

In the countryside, bakers continued to make bread using dough leavened with naturally fermented sponges (sourdoughs) or by the prefermented dough method. However, with rare exceptions, the production of these types of labor-intensive breads was to remain limited, because standardized pricing forbade selling them at a price that provided for the amount of additional labor involved. This was the case with the *pains pliés* (folded breads) of Brittany, the *couronnes fendues* (split crown loaves) or so-called *rosaces* (rose window) loaves, as well as the *pains fendus* (plain split loaves) and the *fendus-tordus* (split and twisted loaves) of southwest France—along with many others.

In some regions, these breads were made from white flour, produced by either stone milling or roller milling. In both cases, the lower grade flours, the second grind (screenings), and bran were carefully reserved for livestock feed.

In other areas, the various breads were made from white rye flour, mixed with wheat flour. This was done in Brittany, the central parts of the country, and sometimes in the higher parts of the alpine areas, where bread is still made in round loaves. At that time only a few types of regional breads were being produced. They required somewhat unusual procedures and had very distinctive flavor characteristics, since they were made from doughs that had been naturally leavened. These included breads from Beaucaire, Aix, and *pain brié* from Normandy.[13]

In the cities, the straight dough method rapidly became more important throughout the 1930s, and almost entirely replaced the "*poolish*." With a few notable exceptions, the straight dough system also replaced the use of the naturally fermented sponge and the prefermented dough method.

There were also changes and improvements to the procedures used in making bread:

- conventional (low speed) dough mixing assured that the fully mixed dough would be only slightly oxidized and not spoiled by overmixing;
- a first fermentation stage lasting from 3 to 4 hours resulted in a slow and natural maturation of the dough, contributing to the production of a significant amount of organic acids. These organic acids aid in the formation of a rich flavor in the crumb, and ensured that the baked loaves would be attractive in appearance, as well as having good shelf life.

The crumb was cream colored, which was the original and most desirable tint. Manual weighing and shaping helped to make the crumb well structured, with good cell formation and excellent eating qualities.

Bakers also made up specially formed breads from these doughs, such as *pains fendus* (split top loaves) and 2-kg split-top crown loaves or *couronnes fendues*. These had a well-merited reputation, but their production was abandoned from 1937 on because of tax problems bakers experienced regarding their production.

The straight dough method is capable of yielding bread that is excellent in all respects, provided the rules of the art are fully respected, and may be used to make a variety of products, including 2-pound loaves, medium-sized *bâtard* loaves, baguettes, and hard rolls. The straight dough method can produce goods equal to those made in the past—or that at least approach them very closely in quality. This may be done by compensating for the lack of flexibility imposed by technical progress (i.e., mechanization) and the rapid production rate by adding a prefermented culture—in the form of either a Polish sponge or prefermented dough.

Thus, it is still possible, through the choice of a naturally leavened sponge, a prefermented dough, or the use of ordinary baker's yeast, to produce breads that are both attractive and good tasting. Breads made by any of these methods will have their own distinctive characteristics and can legitimately claim their proper place in the world of gastronomy.

I should add that other types of breads may be added to the list of those that are made by one of the two methods described above, and help to solve the gastronomic problem of matching the proper breads to other foods. These include traditional "crown" loaves, split loaves, unmolded round country bread, and specialty breads such as yeast-raised sweet doughs, country loaves, 100% rye and mixed wheat and rye breads, nut breads, and Parisian brioches.

Except in the case of bread made from a natural sponge, which requires a little rye flour with a 1.2% ash content, the flour used for these products is always type 55 bread flour, with no addition of bean flour or ascorbic acid, and excluding all high extraction/high ash "gray" flour. It is this high extraction "gray" flour that gives to flour and bread the metallic taste and odor of ash. Whether one wishes to admit it or not, these flours and breads have a taste and an odor that are not characteristic of good bread, and this must be avoided when searching for harmonies between bread and other foods. Personal taste may have considerable bearing on these choices, and it is always good to respect the choices made by one's neighbor.

As far as I am concerned, breads made from high-extraction (85%) gray flour, reconstituted whole-wheat flour, and flour with added bran might all have a certain nutritional or dietary interest for some persons. However, all of these are far from possessing the authentic flavor of bread made from the flour that results from milling the entire white portion of the wheat kernel, and which is, as a result impregnated with much of the wheat germ oil. Furthermore, it is in the proper nature or scheme of things to separate the bran envelope from the wheat berry in order to obtain flour, just as we remove the peel from citrus fruits, and in most cases, from seed-bearing fruits, from potatoes, from carrots, and even nuts and eggs from their shell.

If consumers prefer breads made from whole-grain flours, it is because the flours and breads have some type of taste. Since 1960, in the countryside as well as in the city, consumers have known only yeast-leavened traditional breads that were generally lacking in taste, and the reasons are not limited to the baking characteristics of type 55 bread flour.

The taste of bread is also influenced by a number of other factors, some of which include the following:

- The presence or absence of water vapor in the oven at oven loading affects the flavor. Adequate water vapor (or steam) produces a golden crust with a lightly glazed appearance, while lack of water vapor results in loaves with a gray crust.
- Variations in the degree of baking can produce breads with pale crusts, normal golden crusts, or darker crusts, and the degree of dextrinization (sugar conversion and browning) will have a major effect on bread flavor.
- Differences in bake time affect the thickness of crusts, which influences both bread taste and eating qualities.
- For certain specialty breads, the amount of flour dusting can also have an effect. A layer of dusting flour that is too thick will serve as an insulator and will partially arrest sugar and protein browning. When the flour coating is excessive, the result is a less flavorful crust, while a very light layer of dusting flour may improve crust flavor.

COMPARING BREAD WITH OTHER FOODS

It might be said that particular "harmonies" may exist between a particular bread and a certain food.

We have the choice between bread made with a natural leavening sponge, bread fermented with baker's yeast, and sometimes some type of specialty bread that is given distinctive character by special loaf forming or production techniques. But actually, when breads are high in quality and are good tasting, I like all of them.

Breads made from a naturally leavened sponge or from a prefermented dough have a slightly astringent "country-style" flavor that ranges from slightly to markedly acid. Although this flavor is enticing, it has relatively limited degree of compatibility with most other foods.

But bread leavened with a reasonable amount of yeast,[14] and that has not been overoxidized and denatured by overly intensive mixing, will benefit from a long fermentation and an appropriate degree of dough maturity. It will possess a more delicate flavor closer to that of wheat—with a hint of hazelnuts—and will be compatible to a considerably greater degree with a much larger number of foods.

To be more precise concerning the compatibility between bread and various foods, I would advise that naturally fermented bread be used to accompany the following:

- Strong-tasting *hors-d'oeuvre*, heavy dishes, or cheeses, including meat-based *hors-d'oeuvre* such as raw ham, smoked ham, cured dry sausage, fresh sausage, cooked garlic sausage, *pâtés de foie* made from pork or duck, *gratons*,[15] *rillons*,[16] smoked salmon, herring or anchovy filets, highly seasoned fish pâtés, and smoked eels;
- Wild game dishes, wine-based venison dishes, lentils with salted pork, rabbit in mustard sauce, *rognons sautés au vin* (kidneys sautéed in wine), *coq au vin*, grilled pig's feet, *boudin* or blood sausage, *andouillettes*,[17] tripe dishes, braised beef, beef *bourguignon*, various grilled meats, *charbonnée du boulanger*,[18] beef ribs, *entrecôte* (flank steak) in bordelaise sauce, snails, *cassoulet*,[19] *bouillabaisse*,[20] *brandade de morue*,[21] fish oven-baked in white wine, grilled sardines, grilled lobster, *oeufs en meurette*,[22] grilled red mullet, *cèpes persillade*,[23] *pommes de terre boulangère*,[24] dandelion salad, bacon salad, stuffed cabbage, fried breaded eggplant;
- Roquefort cheese, blue cheeses from the Causses or Auvergne, whole milk *époisse* cheese from Burgundy,[25] Muenster cheese, mature goat cheeses including those from the Pyrenées, and double-cream salted cheese;
- Snacks with dressed pork, sausage, or smoked salmon.

Naturally fermented or "sourdough" breads are indispensable for use in soups, such as onion, cabbage, or pumpkin, or to serve as a trencher beneath a roast, a brace of quail, or a thrush, and for making fried croutons. These breads are also good for making toast for breakfast or for an accompaniment to goose liver *pâté*.

I should like to add that when liquid is poured over it, naturally leavened bread soaks it up without becoming excessively swollen, keeps its flavor, and remains agreeable to chew. On the other hand, yeast-leavened bread swells to the point that it disintegrates, changes into mush, and becomes disagreeable and displeasing in the mouth.

It is interesting to recall the experimental work of M. Fleurent, former professor at the National Institute of Agronomy. Around the turn of the century he became interested in the chemistry of wheat and bread and in comparing the liquid absorbing capacity of breads, he recorded results such as those shown in Table 15–3.

In discussing the association between various foods and breads leavened with baker's yeast, I will make a distinction between (1) long loaves and baguettes, and (2) lightly flour-dusted split-top breads, horseshoe-shaped loaves, and split-top crown loaves, even though all of these are made from the same dough.

When they have been produced with a reasonable amount of leavening and correctly baked with a creamy white crumb, long loaves or baguettes are perfect accompaniments to the *hors d'oeuvres*, dishes, and cheeses listed below:

Table 15–3 Moisture Absorbing Capacity of Two Bread Types by Leavening Method

Bread Type	Liquid Absorption	Temperature	Liquid Absorption	Temperature
Natural sponge	95.4 g (3.37 oz)	20°C / 68°F	131.8 g (4.65 oz)	38°C / 100.4°F
Baker's yeast	183.7 g (6.48 oz)	20°C / 68°F	215.3 g (7.59 oz)	38°C / 100.4°F

- Cooked ham, potted goose or duck, pork fritters, poultry liver mousses, strongly flavored fish terrines, raw vegetables such as grated carrots, beetroot salad, mackerel in white wine, sardines or tuna packed in oil;
- Roasts of pork, veal, or lamb, scallops in cream sauce, meat dumplings, poultry (roasted or sautéed chicken, poached chicken in white sauce, chicken with rice), river fish, fish filets in cream sauce, veal sweetbreads, pigeon dishes, potato purees, potato croquettes, carrot purees, green peas, green beans, creamed spinach, cauliflower, cooked endives, various cooked salads, green lettuce salads, corn salad,[26] endive salad;
- Camembert cheese, Brie cheese, lightly aged goat cheeses, hard Cantal cheese, hard Emmental cheese(from Switzerland), Comté cheese(from the French Juras), cooked *pâtés* (one may almost consider them to be cheeses),[27] white cheese, and good fresh butter.

Lightly flour-dusted and well-baked *pains fendus* (split-top loaves), such as the *fendu simple* or *fendu tordu* (common or twisted split top loaf), *couronne fendue* (split crown), *fer à cheval* (split horseshoe), and even the *couronne lyonnaise*[28] (Lyon crown) "harmonize" perfectly with

- *Pâtés* baked in a crust, sheep's trotters, duck in orange sauce, shoulder or leg of mutton, filet of beef, roast beef, stew meats, potted goose or duck, frog's legs, Burgundy or Savoy style fondue, fresh shelled beans, cooked mutton, Provençal tomatoes,
- The cheeses of Cantal, Salers, or Laguiole, blue cheese from Bresse, Livarot soft paste cheese, *Fourme d'Ambert* cheese, and good fresh butter again!

In discussing other categories of bread, I find that good rye bread with a generous "smear" of sweet butter is suitable to accompany shellfish, smoked ham or salmon—and in the morning, it is good with butter, covered with a good layer of orange marmalade, and accompanied by a good café au lait.

Rye bread with nuts, which always literally melts in the mouth, may accompany the majority of cheeses and green salads.

Finally, a good brioche dough that is just slightly sweet will be the perfect accompaniment to a good raw cured Lyon sausage. A brioche made from the same dough and grilled or toasted will be an excellent companion to the richness of *foie gras*.[29]

In today's wide range of products, certain specialty breads may find a place among sourdoughs and traditional yeast-leavened breads. These specialty products are breads with intermediate characteristics and are well adapted to harmonize with foods from one or another of the families mentioned here.

However, when authentic French bread has been produced in a manner that gives it the taste and aroma that it ought to possess, there is really very little room for improvement. With rare exceptions, the much-vaunted virtues of innovations and new products are found to be only the shallowest delusions.

Such are the opinions that I feel I must express on this vast subject. I consider them to be only a modest personal contribution and certainly do not feel that they are the only truths. I would say, very simply, that they are my truths.

REFERENCES

1. In any discussion of flour in this chapter, the reader should understand that the reference is to wheat flour unless otherwise specified.
2. It is not universally true that nutrition researchers scorn white bread and flour. Research conducted by James W. Anderson and Susan Riddel-Lawrence at the Metabolic Research Group, University of Kentucky and the V.A. Medical Center (*American Journal of Clinical Nutrition*, 1991; 54: 836–40) pointed out that refined, wheat-based bakery products (i.e., made from white flour) are useful in providing significant amounts of soluble fiber in the diet. This can be an effective dietary aid in reducing serum cholesterol, and thus in reducing the risk for coronary heart disease.
3. Trace mineral loss due to the binding effects of phytic acid seems to be a significant factor primarily among those populations that consume very large amounts of wholemeal or high extraction breads. J.G. Reinhold reported in *Ecology of Food and Nutrition* (1972, 1 (3):187–192) that high-extraction rate flours and the omission of leavening and fermentation seemed to contribute to mineral deficiency problems among villagers who consumed bread as their primary source of nutrition. The same author reported in the *American Journal of Clinical Nutrition* (1971, 24 (10):1204–1206) that there was up to two times the amount of phytic acid in unleavened homemade bread in Iranian villages than in leavened bread produced in city bakeries. He points out that the high intake of phytate by vil-

lage dwellers may explain their zinc deficiency problems, since individuals who consumed bread produced in city bakeries did not have this problem.

4. Excess consumption of phytic acid continues to be a matter of concern among some nutritional researchers, even as recently as 1999. An article in *Voedingsmiddelentechnologie* (1997, 30 (5):11, 13) notes that phytic acid in bread, especially whole-meal breads, may bind some trace metals and affect their bio-availability. However, the authors also state that it is possible to effect a reduction of phytic acid content of whole-meal breads to levels approaching that of white bread through sourdough fermentation, although phytic acid levels in rye bread were still appreciably higher.

5. Recent work reported in the journal *Advances in Food Sciences* (1998, 20, (5/6):181–189) by Harinder, Tiwana, and Kaur notes that roller milling of flour, in combination with adequate fermentation and the use of additional lactic acid to adjust pH, makes it possible to produce a whole-meal loaf with a significantly lower phytic acid content.

6. This same question should be asked in reference to older adults, who are a growing component of the total U.S. population.

7. The same question might also be asked regarding U.S. consumers.

8. Similar anti-bread campaigns have occurred in the United States from time to time since that same period.

9. Many of the same conditions applied in North America, both during and following World War I and again during the World War II era. Many types of nonwheat flours were used in commercial bread production during those difficult years, and the quality of the bread available to the American consumer during that time was therefore highly variable.

10. The desirability of this increase has in recent years gained support by the development of the U.S. Nutrition Pyramid and Canadian Nutrition Rainbow, both of which stress the importance of bread and other grain products in a healthful diet.

11. By this Calvel means flour that is free of fava or soybean flour, or of other baking additives.

12. *Le pain qu'est-ce?*

    Le pain, c'est le blé, que l'on sème,
    et c'est le grain, qui sous la terre, germe.
    Le pain, c'est la plante qui lève
    et l'épi qui, au printemps émerge.
    Le pain, c'est l'épi qui fleurit et la fleur fécondée,
    et le grain qui mûrit.
    Le pain, c'est la moisson,
    c'est le blé récolté, c'est le grain ensilé.
    Le pain, c'est le blé au moulin,
    c'est le grain, de l'ivraie, séparé et c'est le blé broyé.
    Le pain, c'est le blé transformé, la mouture blutée et,
    la farine extraite, au fournil transportée.
    Le pain, c'est le boulanger au travail,
    le pétrin en action; c'est, de ferments
    enrichie, la pâte élaborée.
    Le pain, c'est une bonne pâte,
    malaxée sans excès, méthodiquement fermentée.
    Le pain, c'est la pâte, en douceur divisée et pesée,
    ainsi que les pâtons, avec ménagement, tournés.
    Le pain, c'est des pâtons enfournés, correctement levés qui,
    pour finir, seront bien cuits, aussi.
    Le pain, c'est toutes ces astreintes...
    des blés de qualité, des meuniers avertis,
    des boulangers surtout, motivés et capables
    et goûtant bien le pain, par eux, élaboré.
    —Le pain, le bon pain, de pur froment panifié...
    c'est ainsi!—c'est d'abord la nature et puis, ce
    sont des hommes.
    Des hommes, respectant la nature des choses
    et les règles de l'art, attachées, au métier.
    Le pain, c'est tout cela, c'est cela son histoire;
    et c'est, n'en doutons pas, son avenir aussi...

    8 septembre 1987. R. Calvel

13. Brié derives from the apparatus used to mix the stiff dough for this type of bread, which could not be mixed by hand. Because it was mixed with a *brie* it became known as *pain brié*.

14. Calvel means without using an excessive amount of yeast.

15. *Gratons* or *grattons* refer to several similar types of pork or goose *hors-d'oeuvre* generally made by long-cooking small pieces of lean meat in melted fat until tender, then browning them. They are generally eaten as a cold or room temperature appetizer. Numerous quite different regional variations exist, as noted in the *Larousse gastronomique* and other culinary reference works.

16. *Rillons* are cubes of fresh pork belly that are marinated in salt, then simmered in lard until tender, then browned. A specialty of the Loire region. The Touraine version is sometimes coated with a little caramel.

17. *Andouillettes* are generally sausages made from pork chitterlings and seasoned with shallots and mustard. The Cambrai version is made from veal. There are many other regional variations.

18. *Charbonnée au boulanger* is pork cubes braised with red wine. Made in various regions.

19. *Cassoulet* is a white bean stew, brought into the realm of *haute cuisine* by the addition of *confit d'oie*, braised pork, and sausage. Originally a specialty of Provence (especially Toulouse, Carcassonne, and Castelnaudary), it is now widely appreciated throughout France.

20. *Bouillabaisse* is a dish composed of fish (especially rockfish) boiled with herbs, olive oil, and spices. To it are often added various shellfish, and perhaps other fish of the northern Mediterranean region. In Marseilles it is served with a special bread called *marette*. It is a speicalty of Provence, especially the Marseilles area.

21. *Brandade de morue* is a mousse made from salt cod, olive oil, and milk. Originally from Provence.

22. *Oeufs en meurette* are eggs poached in red wine and served in a red wine sauce, a specialty of Burgundy.

23. *Cèpes persillade* consists of cèpe mushrooms sautéed in garlic and parsley. This is a dish from the southwest of France.

24. *Pommes de terre boulangère* is made with sliced potatoes braised in the oven with stock and sliced onions.

25. This is a soft, ripened cheese made from cow's milk, with a very strong flavor.

26. Also called lamb's lettuce or corn lettuce, not to be confused with maize.

27. Here Calvel seems to refer to processed cheeses, since Emmenthal and Comté are both processed.

28. The *couronne lyonnaise* is not a split loaf.

29. *Foie gras* is literally liver from specially fattened geese, but not necessarily processed into a *pâté*.

Selected Works of Professor Raymond Calvel

Compiled by Ronald L. Wirtz, PhD

Note: Nearly all of these articles are in French. This is by no means a complete bibliography of Professor Calvel's publications. However, the number and usefulness of the articles cited here are indicative of his influence in the world of baking—not only in France, but everywhere in the world where French-type hearth breads and related products are made.

ARTICLES

Acide ascorbique et pain français [Ascorbic acid and French bread]. *Le boulanger-pâtissier.* 1989;58: 24–26.

L'action du sel sur les pâtes fermentées et sur les caractéristiques du produit fabriqué [The action of salt on fermented doughs and on the characteristics of the finished product]. *Le boulanger-pâtissier.* 1988;57:14–17.

La baguette 85: la qualité retrouvée [The baguette in 1985: quality rediscovered]. *Le boulanger-pâtissier.* 1985;54:14–16.

La baguette 85: la qualité retrouvée (suite) [The baguette in 1985: quality rediscovered (continued)]. *Le boulanger-pâtissier.* 1985;54:12–13.

Des baguettes en Bade-Wurtemberg [Baguettes in Bad-Wurtemberg]. *Fidèles au bon pain: Bulletin de liaison de l'Amicale des Anciens Élèves et des Amis du Professeur CALVEL.* 1994; No.12:6–7.

Le blé, le pain, la France [Wheat, bread, and France]. *Fidèles au bon pain: Bulletin de liaison de l'Amicale des Anciens Élèves et des Amis du Professeur CALVEL.* 1994: No. 11:20–21, 23.

Le bon pain français aux États-Unis [Good French bread in the United States]. *Le boulanger-pâtissier.* 1988;56:18–19.

Le bon pain français aux États-Unis (suite) [Good French bread in the United States (continued)]. *Le boulanger-pâtissier.* 1988;56:20–21.

La boulangerie au Japon [Baking in Japan]. *Le boulanger-pâtissier.* 1987;56:23, 36.

Brioches régionales et étrangères: le panettone [Regional and foreign brioches: the Panettone]. *Le boulanger-pâtissier.* 1984;53:14–16.

Brioches régionales et étrangères: la fouace de Rodez, la pogne de Romans [Regional and foreign brioches: the fouace of Rodez, the pogne of Romans]. *Le boulanger-pâtissier.* 1984;53:17–21.

Brioches régionales et étrangères: le gâteau des rois en pâte levée sucrée, la fouace des Rameaux [Regional and foreign brioches: the Three Kings cake in leavened sweet dough and the Palm Sunday fouace]. *Le boulanger-pâtissier.* 1984;53:26–28.

Brioches régionales et étrangères: la mouna, la pompe des rois, le gibassier, le pain dulce argentin, la brioche brésilienne [Regional and foreign brioches]. *Le boulanger-pâtissier.* 1984;53:17–21.

Conservation des pâtes au froid et réseau glutineux [Preservation of doughs under freezing and the gluten structure]. *Le boulanger-pâtissier.* 1985;54:14–16.

Cuisson du pain dans le four à bois [Baking bread in wood-fired ovens]. *Le boulanger-pâtissier.* 1981;50:27–29.

De l'usage des farines nord-américaines dans la production de pains spéciaux [The use of North American flours in the production of specialty breads]. *Le boulanger-pâtissier.* 1987;56:22–23.

De l'usage des farines nord-américaines dans la production de pains spéciaux (suite) [The use of North American flours in the production of specialty breads (continued)]. *Le boulanger-pâtissier.* 1987;56:17–19.

Dévelopement de la qualité en pain français [Development of quality in French bread]. *Bull. Anciens Élèves École Franc. Meunerie* 1973;254:59–71.

Effets de l'autolyse naturelle de la pâte en panification [Effects of natural dough autolysis in breadmaking]. *Bull. Anciens Élèves École Franc. Meunerie* 1974;264:288–97.

L'élaboration du pain vite et bien [Making bread well and quickly]. *Fidèles au bon pain: Bulletin de liaison de l'Amicale des Anciens Élèves et des Amis du Professeur CALVEL.* 1994;No.12:16–17.

L'emploi du gluten de froment en panification et dans les productions annexes [The use of gluten in breadmaking and in related products]. *Le boulanger-pâtissier.* 1979;47:21–23.

Évolution, pendant 24 heures, du pH d'une pâte fermentée ensemencée à la levure biologique de boulangerie [24-hour development of the pH of a yeast-based bakery starter sponge]. *Le boulanger-pâtissier.* 1985;54:11–15.

Examens analytiques des farines et valeur boulangère [Analytical testing of bakery flours]. *Le boulanger-pâtissier.* 1982;51:24–27.

Examens analytiques des farines et valeur boulangère [Analytical testing of bakery flours]. *Le boulanger-pâtissier.* 1982;51:17–20.

La fabrication des baguettes et des petits pains viennois [Production of baguettes and small Vienna rolls]. *Le boulanger-pâtissier.* 1981;50:19–21.

La fabrication du croissant (suite) [Croissant production (continued)]. *Le boulanger-pâtissier.* 1984;53:9–11.

La fabrication du croissant: le pétrissage du croissant avec repos-autolyse [Croissant production: mixing and autolysis rest period]. *Le boulanger-pâtissier.* 1984;53:11–14.

Fermentation et panification sur levain naturel [Fermentation and breadmaking with natural sponge starter]. *Le boulanger-pâtissier.* 1980;49:25–26.

Fermentation et panification sur levain naturel (suite) [Fermentation and breadmaking with natural sponge starter (continued)]. *Le boulanger-pâtissier.* 1980;49:22–25.

Fermentation et panification sur levain naturel (suite) [Fermentation and breadmaking with natural sponge starter (continued)]. *Le boulanger-pâtissier.* 1980;49:16–19.

Fermentation et panification sur levain naturel (suite) [Fermentation and breadmaking with natural sponge starter (continued)]. *Le boulanger-pâtissier.* 1980;49:30–34.

Flaveur et pain [Flavor and bread]. *Le boulanger-pâtissier.* 1980;49:12–14. Introduction à l'article La flaveur des aliments et le consommateur par Mme Caul de Kansas State University à Manhattan, KS, É-U.

Force et valeur boulangère des farines: force de la pâte et qualité du pain [Flour strength and breadmaking value of flours: dough strength and bread quality]. *Le boulanger-pâtissier.* 1984;53:18–21.

Force et valeur boulangère des farines: force de la pâte et qualité du pain (suite) [Flour strength and breadmaking value of flours: dough strength and bread quality (continued)]. *Le boulanger-pâtissier.* 1984;53:11–14.

L'influence du pétrissage sur l'évolution de la panification et de la qualité du pain [The influence of mixing on the evolution of breadmaking and the quality of bread]. *Le boulanger-pâtissier.* 1979;43:30–32.

L'influence du pétrissage sur l'évolution de la panification et de la qualité du pain [The influence of mixing on the evolution of breadmaking and the quality of bread]. *Le boulanger-pâtissier.* 1979;43:25–26.

Le Japon, la Corée [Japan and Korea]. *Fidèles au bon pain: Bulletin de liaison de l'Amicale des Anciens Élèves et des Amis du Professeur CALVEL.* 1994;No.11:25–26.

Les matériels accessoires à la cuisson: moules, plaques, filets et démoulage [Accessory equipment in baking: bread pans, sheets, nets, and depanning]. *Le boulanger-pâtissier.* 1984;53:14–16.

Les matériels accessoires à la cuisson: moules, plaques, filets et démoulage (suite) [Accessory equipment in baking: bread pans, sheets, nets, and depanning]. *Le boulanger-pâtissier.* 1984;53:14–15.

La mécanisation du travail des pâtes et ses conséquences sur la qualité du pain [Mechanisation of dough handling and kneading and consequences on bread quality] *Le boulanger-pâtissier.* 1980;49:17–21.

La mécanisation du travail des pâtes et ses conséquences sur la qualité du pain (suite) [Mechanisation of dough handling and kneading and consequences on bread quality (continued)]. *Le boulanger-pâtissier.* 1980;49:14–15.

Le pain au son [Bran bread]. *Le boulanger-pâtissier.* 1982;51:31–33.

Le pain complet [Whole-grain bread]. *Le boulanger-pâtissier.* 1982;51:16–20.

Le pain de seigle: levain naturel, pâte fermentée [Rye bread: natural sponge, fermented dough]. *Le boulanger-pâtissier.* 1989;58:12–13.

Pain français, croissants et brioches régionales au Japon [French bread, croissants, and regional brioches in Japan]. *Le boulanger-pâtissier.* 1988;57:43–45.

Pain français, croissants et brioches régionales au Japon (suite) [French bread, croissants, and regional brioches in Japan (continued)]. *Le boulanger-pâtissier.* 1988;57:32–34.

Le pain français, de l'Extrême-Orient à la Californie: vulgarisation et formules [French bread, from the Far East to California: popularization and formulas]. *Fidèles au bon pain: Bulletin de liaison de l'Amicale des Anciens Élèves et des Amis du Professeur CALVEL.* 1992;No.8:19–28.

Le pain rustique farine pur froment [Rustic breads from 100% wheat flour]. *Le boulanger-pâtissier.* 1985;54:11–13.

Le pain rustique pur froment [Rustic breads from 100% wheat flour]. *Le boulanger-pâtissier.* 1983;52:21–23.

Panifier vite et bien [Making bread quickly and well]. *Le boulanger-pâtissier.* 1980;49:14–15.

Le pastis landais ou une brioche rustique [The Landais 'pastis' or a country-style brioche]. *Le boulanger-pâtissier.* 1985;54:9–11.

Les pâtes fermentées surgelées [Frozen bread doughs. par Hubert Maître)]. *Le boulanger-pâtissier.* 1985;54:11–17.

Les pâtes fermentées surgelées (suite) [Frozen bread doughs (continued). par Hubert Maître)]. *Le boulanger-pâtissier.* 1985;54:12–17.

La panification au levain naturel [Breadmaking with a natural starter sponge]. *Fidèles au bon pain: Bulletin de liaison de l'Amicale des Anciens Élèves et des Amis du Professeur CALVEL.* 1988;No.3:25–36.

Les caractéristiques du bon pain français [The characteristics of good French bread]. *Fidèles au bon pain: Bulletin de liaison de l'Amicale des Anciens Élèves et des Amis du Professeur CALVEL.* 1987;No.1:22–27.

Panification rapide de qualité avec apport de pâte fermentée [Rapid production of quality bread with the addition of prefermented sponge]. *Le boulanger-pâtissier.* 1980;49:16–18.

Panification rapide de qualité avec apport de pâte fermentée (suite) [Rapid production of quality bread with the addition of prefermented sponge (continued)]. *Le boulanger-pâtissier.* 1980;49: 16–17.

La panification sur poolish [Breadmaking with Polish-style sponge yeast starter)]. *Le boulanger-pâtissier.* 1989;58:14–15. Note sur James MacGuire au restaurant *Passe–Partout* à Montréal.

Les pains arômatiques [Specialty herb breads]. *Le boulanger-pâtissier.* 1987;56:18–19.

Les pains arômatiques (suite) [Specialty herb breads (continued)]. *Le boulanger-pâtissier.* 1987;56: 18–20.

Pains au gluten et produits hyperprotéiques et hypocaloriques [Gluten breads and high-protein and low-calorie products]. *Le boulanger-pâtissier.* 1982;51:10–12.

Pains au gluten et produits hyperprotéiques et hypocaloriques (suite) [Gluten breads and high-protein and low-calorie products]. *Le boulanger-pâtissier.* 1982;51:19–20.

Le pétrissage avec bras sous forme de spirale [Kneading with spiral mixers]. *Le boulanger-pâtissier.* 1988;56:22–23.

Principles and techniques of quality French bread. Notes en anglais du colloque sur la panification française donné à Kansas State University, Manhattan, KS, É.-U., du 8 au 11 novembre, 1987. 34 p

La production des brioches [The production of brioches]. *Le boulanger-pâtissier.* 1983;52:18–20.

La production du pain de campagne [The production of country-style rustic) breads]. *Le boulanger-pâtissier.* 1986;55:16–19.

La production du pain de campagne [The production of country-style rustic breads]. *Le boulanger-pâtissier.* 1981;50:26–30.

La production du pain de campagne (suite) [The production of country-style rustic breads (continued)]. *Le boulanger-pâtissier.* 1981;50:20–22.

La production du pain de campagne (suite) [The production of country-style rustic breads (continued)]. *Le boulanger-pâtissier.* 1986;55:14–18.

La production du pain de mie [The production of white pan bread]. *Le boulanger-pâtissier.* 1981;50:16–18.

La production du pain de mie [The production of white pan bread]. *Le boulanger-pâtissier.* 1981;50:23–25.

La production du pain de seigle [The production of rye bread]. *Le boulanger-pâtissier.* 1981;50:26–28.

La production du pain de seigle [The production of rye bread]. *Le boulanger-pâtissier.* 1981;50:21–23.

La production du pain français au Japon [Making French bread in Japan]. *Le boulanger-pâtissier.* 1986;55:13–17.

La production du pain français au Japon (suite) [Making French bread in Japan (continued) par Kiochi Fukumori]. *Le boulanger-pâtissier.* 1986;55:13–19.

La production d'un pain de mie de qualité [Making quality white pan bread]. *Le boulanger-pâtissier.* 1988;57:12–14.

La production d'un pain de mie de qualité (suite) [Making quality white pan bread (continued)]. *Le boulanger-pâtissier.* 1988;57:16–19.

Les produits auxiliaires de panification [Production of other bakery goods]. *Le boulanger-pâtissier.* 1987;56:19–20.

La qualité des blés dans le monde, face aux exigences de l'élaboration des baguettes [World wheat quality, in reference to requirements for baguette production]. *Fidèles au bon pain: Bulletin de liaison de l'Amicale des Anciens Élèves et des Amis du Professeur CALVEL.* 1994;No.11:33–37.

La qualité du pain: réponse aux exigences des consommateurs [Bread quality: response to consumer requirements]. *Le boulanger-pâtissier.* 1982;51:20–22.

La qualité du pain: réponse aux exigences des consommateurs (suite) [Bread quality: response to consumer requirements (continued)]. *Le boulanger-pâtissier.* 1982;51:22–25.

Un quatrième type de pâte – la pâte surhydratée ou la pattemouille *Fidèles au bon pain: Bulletin de liaison de l'Amicale des Anciens Élèves et des Amis du Professeur CALVEL.* 1988;No.17.

Regard sur l'histoire: l'emploi de la levure dans la fermentation panaire [A historical view of the use of yeast in breadmaking]. *Fidèles au bon pain: Bulletin de liaison de l'Amicale des Anciens Élèves et des Amis du Professeur CALVEL.* 1994;No.12:29–30.

La surgélation en panification, ou la médaille et son revers [Freezing in breadmaking, pro and con]. *Le boulanger-pâtissier.* 1988;57:15–18.

La surhydration de la pâte et la qualité de la croûte de pain [Superhydration of dough and the quality of bread crust]. *Le boulanger-pâtissier.* 1987;56:8–10.

Viennoiserie et pâtes levées-sucrées [Vienna breads and sweet doughs]. *Le boulanger-pâtissier.* 1983;52:13–15.

Viennoiserie et pâtes levées-sucrées (suite) [Vienna breads and sweet doughs (continued)]. *Le boulanger-pâtissier.* 1983;52:12–15.

Viennoiserie et pâtes levées-sucrées (fin) [Vienna breads and sweet doughs (end)]. *Le boulanger-pâtissier.* 1983;52:14–17.

BOOKS

Le goût du pain: comment le préserver, comment le retrouver [The taste of bread: how to keep it, how to rediscover it]. Paris: Éditions Jérôme Villette; 1990.

La boulangerie moderne [The modern bakery]. 3rd ed. Paris: Éditions Eyrolles; 1962.

Index

A

Absorption, xii, 11, 36, 81, 185, 186, 193
Acids, organic, 16, 30, 38, 42, 43, 44, 49–54, 56, 62, 85, 94, 98, 143, 146, 191
American Institute of Baking, ix, 6, 8, 28, 37
Amylase, 15, 17, 18, 35, 37, 38, 78, 81, 99
Applied Baking Technology, 37
L'Art du boulanger (1667), 48
Artisan, vii, 4, 7, 17, 19, 22, 43, 58, 83, 137, 139, 140
Alveogram, 10, 11, 13, 132
Alveograph, 11, 13, 21, 23, 35, 111, 128
American Association of Cereal Chemists (AACC), 14
Ascorbic acid, 15, 16, 18, 31, 33, 34, 35, 36, 39, 77, 96, 99, 128, 131, 146
Autolysis of dough, 31, 34, 40, 91, 93, 94, 99, 103, 105, 143, 152
Azodicarbonomide, 18, 22, 48

B

Bacillus mesentericus, 84
Baker's percent, xi
Baker's yeast:, 19–20, 22, 38, 40, 41, 42, 45, 47, 48, 49, 50, 51, 56, 89, 91, 92, 94, 95–96, 98, 107, 116, 190, 191, 193. See also *Saccharomyces cerevisiae*
Baking, 67, 69, 71–73
Barrier, Charles, vii
Benzoyl peroxide, 18, 22
Biga, 48
Boulanger-Pâtissier, Le, ix, 14

Boulangerie Française, La, xiii
Boulangerie Moderne, La, ix
Bour, D., 187
Bread infection. *See* Spoilage
Bread recipes. *See* end of index.
Brioche, xi, xii, xiii, 9, 10, 96, 114–116, 129, 141, 149–156, 192, 194
    conventional Paris style, 149–150
    formula, procedures, and leavening, 150, 152
Brioche recipes. *See* end of index.
Bulletin des Anciens Elèves et des Amis du Professeur Calvel, xiv, 29

C

Calcium propionate, 16, 17, 84
Canadian Wheat Board, 5
Caramelization, 23, 57, 72, 76, 78, 191
Carotenoid pigments, xiv, 18, 32, 35, 36
Charlelègue, M., 35
Chef, 22, 42, 90, 91, 92, 93, 106, 158, 160, 165, 168, 169, 170, 174
Chiron, Hubert, vii, ix, 13, 74, 102
Chlorine, 18, 22
Chopin alveographe. *See* alveograph
Citric acid, 16, 17, 22, 107
Couche, 60
Croissant, xii, xiii, color plate 10, 10, 98, 141–148, 149, 156
    origin of, 141
    pain au chocolate, 148–149

201

production, 141–148
  cutting, 145
  fermentation, 143
  lamination, 143, 144, 145
  unbaked, frozen, 146
  under refrigeration, 146
Croissant recipes. *See* end of index.
Crumb, 78–79
  aroma of, 49, 51, 52, 53, 78, 83–84
  cell structure, 79
  color of, iv, 9, 34, 35, 40, 79
  organic acids in, 50–51
  structure of, 19, 31, 32, , 35, 41, 56, 57, 58, 59, 60, 61, 68, 69, 73, 78, 79, 81, 91, 102, 113, 114, 128, 131, 135, 136, 143, 152, 154, 175, 184, 185, 190, 192
Crust, 67–77
  bubbles as defect, color plate 11
  burning, 73
  color of, color plate 14, 17, 18, 19, 30, 35, 44, 56, 57, 72, 73, 76, 119, 190
  effects of oven steam, 73
  flavor of, 57, 72, 73, 76, 78, 83
  flour dusting, color plate 13, 75–76
  formation, coloration, degree of crust baking, 69
  frozen storage, 77
  ovens used in baking, 67–69
  scaling of, 61, 76–77
  temperature of, 72
Culinary Institute of America, 90
Cutting blade, 70

D

Direct, straight-dough method, 42, 43
Dough, 15–23
  additives, 16–17, 18
  bleaching effect, 8, 18, 19, 22, 27, 30, 35, 36, 37, 54, 152
  characteristics, 27
  cohesiveness, 9, 17, 18, 19, 27, 28, 38, 39, 43, 55, 56, 69, 81, 152
  composition 15–16
  dividing, 58, 101
  elasticity. *See* cohesiveness
  extensibility, 9, 12, 18, 27, 31, 38, 39, 55, 91, 132, 142, 143
  ingredients, influence of, 18–22
  maturation of, 16, 19, 27, 29, 30, 32, 35, 36, 38, 39, 40, 43, 44, 54, 55–63, 79, 90, 94, 152, 155, 191
  "punching down" of, 40, 43, 45, 111, 152, 155
  shaping, 59
    long loaves, 60

  temperature, xi, xii, 27, 28, 30, 37, 39, 40, 44, 48, 51, 52, 61, 69, 78, 91, 93, 97
  weighing, 15, 59
Dough temperature calculation, 28, 37
Dough types, French, 29
Draperon, R., 34, 37
Dupin, P., 187

E

École Nationale Supérieure de Meunerie et des Industries Céréalières. *See* ENSMIC
ENSMIC, xiv
Enzymes, 12, 13, 17, 34, 35, 38, 105, 127, 128
Europain, ix, 45
Extraction, rate of, xiii, 3, 4, 5, 6, 7, 9, 10, 102, 109, 183, 184, 185, 187, 188, 192, 194

F

Faba bean flour, xiv, 15, 16, 17, 18, 30, 34–37, 50
Falling number index, 11–12
Farine de gruau, 9, 10, 29, 103, 105, 116, 128
Fermentation, 38–48
  acid, 21, 41
  alcoholic, xiv, 19, 20, 21, 29, 30, 38, 42, 43, 47, 59, 79, 81, 96, 164, 171, 191
  bulk, xii, 22, 36, 37, 48, 135
  effect on taste, 40–44
  primary, 30, 43, 85, 143
  role of, 38–40
  secondary, 43, 47, 50, 61
  total time, 51
Flour, 3–14
  additives, 18
  aging of, 14, 35, 36,
  ash content, 3, 4, 7, 9, 10, 14, 16, 105, 111
  baking quality, 11, 36, 127, 141,
  blended stream, 5, 7, 10, 14, 17, 102
  characteristics, 4, 7, 11, 14
  classifications,
    France, 4, 9, 10
    United States, 4, 10
  diastatic power, 11, 12, 57
    hyperdiastasticity, 13, 34, 57, 73, 107
    hypodiastasticity, 13, 17, 30, 35, 37, 73
  "green". *See* Flour, aging of
  humidity, 3, 4, 10, 12, 13,
  milling of, 5, 7, 8
  mixtures, 111
  nature of, 3–10

rye, 3, 10, 17, 22, 89, 90, 92, 101, 102, 103, 105, 106, 107, 108, 109, 121, 128, 167, 191, 192,
rye types, 10
straight process, 5, 10
stone ground, 4, 10, 128, 183, 189, 190, 191
type 55, xiii, 4, 9, 10, 13, 16, 79, 84, 92, 102, 104
"W" value, 11, 12
wheat, 3, 6
Freezing, effects of, 55, 61–63, 77
French bread, 89–101
   levain and *levain de pâte*, 89–92
   rustic (country style) with pure-wheat flour, 99–101
   yeast raised (pain courant), 94–99
French National School for Milling and Cereal Processing Industries. *See* ENSMIC

G

Girard, Aimé, 185
Gisslin, Wayne, 37
Gluten/gluten network, 4, 11, 16, 28, 30, 31, 32, 35, 36, 40, 62, 71, 77, 78, 89, 91, 105, 109, 132, 150, 152, 190
Godon, B., 83
Gounelle de Pontanel, H., 185
Grigne, 70
Guilbaud, A., 34, 36, 37, 83
Guilbot, A. *See* Guilbaud, A.

H

Hagberg apparatus, 11, 12
Hydration, xii, 3, 11, 29, 81, 90, 101
   North American flours, 29, 44, 90, 101

I

Ice, used in dough temperature regulation, 28
Institut National de Recherche Agronomique (INRA), 13, 34, 37, 83, 89

J

Japan, xiii, 13, 14, 98, 111

K

Kansas State University, ix, 14, 90, 128
Kansas Wheat Commission, 5
Kline, L. *See* Western Regional Research Laboratory

L

Lactobacillus sanfrancisco, 48
Launay, B., 51
Laws, France, 17, 118
Leavening
   discovery of, 21
   evolution of 45–47
   schematic comparison, 46
   systems of, 39, 42
      differences, 49
      poolish, 39, 42, 43, 45, 48, 49, 50, 51, 52, 53, 54, 56, 57, 82, 94, 96, 98, 116, 121, 191
      prefermented dough, vii, 32, 39, 42, 43, 44, 48, 49, 51, 57, 62, 82, 85, 92, 94, 95, 96, 98, 99, 104,107 108, 120, 121, 123, 143, 147, 148, 150, 152, 156, 170, 176, 191, 192
      sourdough (*levain*), 21, 22, (31), 39, 41, 43, (47), 48, (49), (50), (53), 56, 63, (73), 74, 75, (79), (82), (85), (89), 90, (91), (92), 95, (101), 103, 107, (116), 128, 150, 158, (172), (173), (174), 180, 191, 193, 194, 195
      sponge and dough, 39, 42, 44, 45, 49, 52, 56, 57, 95, 96, 98, 111, 118, 135, 136, 137, 150, 156, 158, 160, 161, 163, 165, 167, 168, 169, 190, 191
      straight dough, 14, 39, 43, 44, 45, 47, 49, 50, 51, 52, 53, 54, 79, 82, 85, 94, 102, 111, 116, 118, 120, 121, 127, 135, 139, 146, 148, 150, 152, 153, 158, 160, 161, 163, 166, 167, 191, 192
Lecithin, 15, 16–17, 22
Lestradet, P., 186, 187
Levain, 41, 89. *See also* sourdough cultures, natural
Levain de pâte, 41, 89. *See also* leavening systems, prefermented dough
Levain-levure, 42. *See also* leavening systems, sponge and dough
Lipoxygenase, xiv, 30, 31, 34, 35, 36, 40, 52, 53
Loaf forming
   effects of, 57
   mechanical versus manual, 57–61
   seam placement, color plate 12
Loaf types
   baguette, xii, xiii, color plate 4, color plate 5, 41, 43, 47, 57, 58, 59, 60, 62, 69, 70, 74, 75, 77, 81, 82, 85, 94, 95, 96, 98, 102, 103, 109, 116, 123, 140, 192, 193
   bâtard, 43, 74, 75, 116, 192
   boule, 72, 103
   boulot, 41, 74,
   brié, 191
   country style, color plate 6, 72, 102
   couronne fendue, 191
   épi, 74

fendu, 71
fendu-tordu, 191
ficelle, 74
levain, color plate 2, color plate 7, 73
marchand de vin, 44, 47, 72,
miche, color plate 3, 103
parisien, 74
petit pain, 74
plié, 191
polka, 75,
pullman, 83
rosace, 72
rustic, color plate 2, 29, 73, 75, 85, 89, 100, 101, 102, 103, 191
sandwich, 83
split loaf, 43
tourte, 103
walnut, color plate 9
whole wheat, color plate 2

M

McMahon, Tom, 4
Mailliard reaction, 23, 57
Malouin, M., 48
Méthode directe. See *leavening systems, straight dough*
Metric system, xi
Mie. See Crumb
Milton-Keynes process, 63. *See also* par-baking process
Mixers, types, 33–34, 40, 41
Mixing, 27–37
  methods of, 32
    improved, color plate 1, 32
    intensive, color plate 1, 32
    traditional, color plate 1, 32
  mixing times, 29, 40, 41
  oxidation, 28, 30, 32–37
Moisture, 80, 193
Mold inhibitors, 84
Molding, loaf. *See* Loaf forming
Mono- and diglycerides, 18

N

National Baking Center, 4
Natural sponge, 38
Naturally fermented sourdough. *See* Sourdough cultures, natural
Nutritional contribution of bread, 185

O

Oven spring effect, 69
Ovens, 67–69
  ash, 7
  electric, 69
  fixed deck, xi, 67, 68, 69
  masonry, 67, 73
  rack, 61, 62, 68, 69, 73
  rotating sole, 45, 67, 69, 101, 134, 140
  tunnel, 69, 84, 131, 134, 135, 139,
  wood-fired, 67, 68, 197
Oxidation, xiv, 1, 3, 14, 17, 22, 23, 27, 28, 29, 30, 31, 32, 33, 35, 38, 39, 40, 41, 43, 44, 48, 49, 50, 53, 61, 72, 75, 76, 77, 78, 79, 81, 82, 94, 98, 101, 102, 134, 158, 190

P

Pain brioché, xi
Pain français, Le, xiii
Pain Français et les productions annexes, Le, xiii
Pain maison, 17
Pain tradition, 17
Pain viennois, 45
Panary fermentation. *See* fermentation, alcoholic; fermentation, acid
Par-baking process, 61–63
Passe-Partout (bakery), ix
Pâte fermentée. *See* Leavening systems, prefermented dough
Per capita consumption, xiv, 188
pH, 12, 22, 55, 56–57, 82, 90, 180, 195
Plat, D., 187
Point, Fernand, xi
Pointage. See fermentation, primary
Poolish, 42. *See also* leavening systems, liquid sponge
Potassium bromate, 18, 22, 48,
Pradini-Jarre, D., 185
Prefermented dough, 44
  base formula, 44
Proof, degree of, color plate 13, 60, 61
Propionic acid, 16, 17, 50, 84
Protein content, 3, 4, 5, 12, 13, 73, 105, 121

Q

Qualities of bread, 183–195
  bread and gastronomy, 190–192
  bread made from stone-ground flour, 189–190

caloric content and bioavailability, 185–188
comparing bread with other foods, 192–193
nutritional value, 183–185
progressive decline of bread consumption in France, 188–189

R

Regional brioches, 158–180
    Argentine sweet bread, 179
    Brazillian brioche, 179
    Fouace de Rodez, 166
    Fouace des Rameaux, 163
    Gâche hearth cake, 161
    Gateau des rois, 163
    Gibbassier, 171–174
    Kugelhopf, 158–160
    La pogne de Romans, 167
    Mouna, 167
    Pompe des rois, 170–171
    Vendée-style, 161
Richard-Molar, D., 72
Robuchon, Joel, vii
Rope. *See Bacillus mesentericus*
Rosada, Didier, 4
Rusks and specialty toasted breads, 131–140
    biscotte courant, 131–137
    breadsticks and grissini, 139–140
    gluten-free breads, 137–139

S

Saccharomyces cerevisiae, 19, 89
Saccharomyces minor, 21
Salt, influence of, 19, 22, 23, 30, 34, 36, 37, 39, 42, 56, 89
Scarification of loaves, color plate 13, 55, 70, 74
Serrage, 58
Shelf life, 83–84
    Slashing/cutting of loaves. *See* Scarification
Sosland Publishing Company, 22
Sourdough cultures, commercial
    active cultures, 21
    deactivated cultures21
    liquid cultures, 21
Sourdough cultures, natural, 21
    building a starter, 89–92
    Levain, 21
    Levain de pâte, 22
Soy flour, xiv, 1, 16, 17, 30, 36, 37, 54, 99, 111, 127, 138, 195

Specialty breads, 102–128
    breads for filling and topping 119–120
    brioche-type, 114
    country style, 102–103, 104
    five grain, 109–111
    gluten, 118–119
    milk rolls, 116
    pain de gruau, 102, 105
    "pistol" rolls, 118
    raisin, 109
    rye, 103, 104, 105–107
    sandwich, 118
    savory and aromatic breads, 120–127
        algae or seaweed, 123
        anise, 121
        apple, 125
        bacon, 124
        cheese, 122–123
        cumin, 122
        fennel, 122
        garlic, 122
        olive, 121
        onion, 122
        poppyseed, 122
        potato, 126
        pumpkin, 124
        sausage, 125–126
        soy, 127
        sunflower oil bread with sesame seeds, 123
        thyme, 122
        wheat germ, 123
        zuchinni, 124
    vienna, 116
    walnut, 108
    white-tinned pullman, 111
    whole grain and bran, 109
Spoilage, 36, 80, 82, 83, 84, 85, 156. *See also* Staling
    molds, 84
    ropy bread, 84–85
Staling
    acceleration of, 77, 81, 82, 85
    consumption, 83
    factors influencing staling, 81–83
    inhibition of, 82, 85
    shelf life, 83–84
    spoilage, 84–85
    storage and staling effects, 80–81
Starters, 21–22, 40–45
Steam, color plate 12, 73, 75
Stone-ground flour, 189–190
Storage, 77, 80
Straight dough with addition of prefermented dough, 44

Strecker reaction, 72
Sugars, 11, 17, 23, 30, 36, 38, 44, 55, 56, 57, 61, 63, 71, 72, 73, 78, 121, 135, 136, 146, 150, 152, 153, 154, 183, 186. *(Sugar as an ingredient is listed in many recipes in addition to these listings.)*
Sugihara. *See* Western Regional Research Laboratory

T

Thermometer, use of, xi, 39, 48,

V

Vague, P., 187
Vigne, P., 187

W

Water
   influence on dough, 18
   mineral content, 18, 22
   temperature, xii, 27, 28, 37
Western Regional Research Laboratory (California), 89
Wheat, 3, 6
   grit flour. See *Farine de gruau*
   hard red spring, 5
   hard red winter, 5
   protein content, 3
   structure of kernel, 6

Y

Yeast, influence of, 19–21, 42–46
Yeast, types of
   baker's, 19
   dry, 20
   granulated yeast, 20
   yeast, 20
Yeast-raised sweet doughs, 141–157
   brioches, 149–156
   chocolate-filled buns from croissant dough, 148, 149
   snail rolls, 149
   traditional croissants, 141–148

Z

Zang, Baron, 116

RECIPES

100% whole wheat bread, 110 (Exhibit 11–9)
20% bran bread, 110, (Exhibit 11–10)
Basic French bread
   with poolish (using four-fifths of formula water, no autolysis), 97 (Exhibit 10–7)
   with poolish (using one-third of formula water, no autolysis), 97 (Exhibit 10–6)
   prefermented dough (with autolysis), 99 (Exhibit 10–9)
   sponge and dough method (no autolysis), 98 (Exhibit 10–8)
   straight dough method with improved mixing, 95 (Exhibit 10–4)
   straight dough method with intensive mixing, 96 (Exhibit 10–5)
   straight dough process with traditional mixing method, 94 (Exhibit 10–3)
Brioche
   cooked Lyon sausage in brioche, 156 (Exhibit 13–7)
   methods
      prefermented dough method, 151 (Exhibit 13–5)
      sponge and dough method, 151 (Exhibit 13–6)
      straight dough method, 150 (Exhibit 13–4)
   regional brioche
      Argentine sweet bread, *pan dulce*, 180 (Exhibit 14–22)
      Brazilian brioche sponge and dough method, 179 (Exhibit 14–21)
      Fouace de Rodez, Rodez hearth cake
         levain de pate, 169 (Exhibit 14–11)
         naturally fermented sponge, 168 (Exhibit 14–10)
         straight dough method, 167 (Exhibit 14–9)
      Fouace des rameaux, Palm Sunday hearth cake
         naturally fermented sponge, 165 (Exhibit 14–7)
         sponge and dough method, 166 (Exhibit 14–8)
      Gâche, 162 (Exhibit 14–4)
      Gibassier, sponge and dough method, 174 (Exhibit 14–16)
      Kings' cake
         sponge and dough method, 164 (Exhibit 14–6)
         straight dough method, 164 (Exhibit 14–5)
      kugelhopf
         sponge and dough method, 159 (Exhibit 14–1)
         straight dough method, 160 (Exhibit 14–2)
      mouna, sponge and dough method, 172 (Exhibit 14–14)
      panettone
         emulsion, 177 (Exhibit 14–19)
         "imitation" pannetone by sponge and dough method, 178 (Exhibit 14–20)

industrial formula, 176 (Exhibit 14–18)
levain for panettone, 175 (Exhibit 14–17)
Pogne de Romans
  naturally prefermented dough, 170 (Exhibit 14–12)
  sponge and dough method, 171 (Exhibit 14–13)
Pompe des rois, Kings' feast cake, 173 (Exhibit 14–15)
Vendée-type, 162 (Exhibit 14–3)
Brioche-type bread (*pain brioché*)
  prefermented dough method, 115 (Exhibit 11–16)
  straight dough method, 115 (Exhibit 11–15)
Country style bread
  prefermented dough method, 104 (Exhibit 11–1)
  sponge and dough method, 104 (Exhibit 11–2)
Croissants
  croissants made with margarine (with autolyis), 142 (Exhibit 13–2)
  croissants with addition of prefermented dough, 142 (Exhibit 13–3)
  traditional croissants made with butter, 142 (Exhibit 13–1)
Gluten bread, 119 (Exhibit 11–21)
Gluten-free bread, 139 (Exhibit 12–7)
Levain de Pâte, 93 (Exhibit 10–2)
Milk rolls, 117 (Exhibit 11–18)
Multigrain bread
  prefermented dough method, 112 (Exhibit 11–12)
  traditional straight dough method, 111 (Exhibit 11–11)
Naturally leavened sponge, 91 (Exhibit 10–1)
Pain de gruau straight dough method with autolysis, 105 (Exhibit 11–3)
"Pistol" rolls, 119 (Exhibit 11–20)
Pizza, *pissaladiere* (dough for filling or topping), 120 (Exhibit 11–22)
Rusks
  breadsticks and grissini, 139 (Exhibit 12–8)
  formulas and procedures, 132 (Exhibit 12–1)
  high-calorie weight-gain rusk with addition of prefermented dough, 133 (Exhibit 12–3)
  Hollund rusks, 137 (Exhibit 12–5)
  low-calorie dietary rusk with addition of fermented dough, 133 (Exhibit 12–2)
  specialty bread for pre-toasting with addition of fermented dough, 136 (Exhibit 12–4)
  Turin-style grissini, straight dough method, 140 (Exhibit 12–9)
  zwieback, sponge and dough method, 138 (Exhibit 12–6)
Rustic bread from pure wheat flour, 100 (Exhibit 10–10)
Rye bread (*pain de seigle*)
  with addition of prefermented dough and baker's yeast, 107 (Exhibit 11–7)
  traditional sourdough rye, 106 (Exhibits 11–4 through 11–6)
Rye walnut bread (mixed wheat and rye with walnuts), 108 (Exhibit 11–8
Sandwich rolls, 118 (Exhibit 11–19)
Savory and aromatic bread
  apple bread, poolish sponge method, 125 (Exhibit 11–29)
  base formula, 122 (Exhibit 11–23)
  potato bread, 126 (Exhibit 11–30)
  pumpkin bread, 125 (Exhibit 11–28)
  smoked bacon bread, 124 (Exhibit 11–26)
  soy bread
    prefermented dough method, 127 (Exhibit 11–33)
    straight dough method, 127 (Exhibit 11–32)
  sunflower oil bread with sesame seeds, 123 (Exhibit 11–24)
  Toulouse sausage bread, prefermented dough method, 126 (Exhibit 11–30)
  wheat germ dietetic bread, 123 (Exhibit 11–25)
  zucchini bread, 124 (Exhibit 11–27)
Tinned sandwich or pullman bread
  prefermented dough method, 112 (Exhibit 11–13)
  straight dough method with added milk, 113 (Exhibit 11–14)
"Vienna" bread and rolls by prefermented dough method, 117 (Exhibit 11–17)

Made in the USA
Lexington, KY
20 March 2015